Power Systems

Electrical power has been the technological foundation of industrial societies for many years. Although the systems designed to provide and apply electrical energy have reached a high degree of maturity, unforeseen problems are constantly encountered, necessitating the design of more efficient and reliable systems based on novel technologies. The book series Power Systems is aimed at providing detailed, accurate and sound technical information about these new developments in electrical power engineering. It includes topics on power generation, storage and transmission as well as electrical machines. The monographs and advanced textbooks in this series address researchers, lecturers, industrial engineers and senior students in electrical engineering.

Power Systems is indexed in Scopus

More information about this series at http://www.springer.com/series/4622

Alicia Triviño-Cabrera ·
José M. González-González ·
José A. Aguado

Wireless Power Transfer for Electric Vehicles: Foundations and Design Approach

 Springer

Alicia Triviño-Cabrera
Escuela de Ingenierías Industriales
University of Malaga
Málaga, Spain

José M. González-González
Escuela de Ingenierías Industriales
University of Malaga
Málaga, Spain

José A. Aguado
Escuela de Ingenierías Industriales
University of Malaga
Málaga, Spain

ISSN 1612-1287   ISSN 1860-4676   (electronic)
Power Systems
ISBN 978-3-030-26705-6   ISBN 978-3-030-26706-3   (eBook)
https://doi.org/10.1007/978-3-030-26706-3

This Springer imprint is published by the registered company Springer Nature Switzerland AG
The registered company address is: Gewerbestrasse 11, 6330 Cham, Switzerland

To my loving sons, Pablo and Dani, for teaching me the true meaning of LIFE.

To my family, specially to my nephew, Pablo, and to the members who are not here, my grandfathers, who support me wherever they are.

To my wife, Patricia, and my son, José Antonio, for being my source of inspiration.

Preface

The scientific foundations of wireless power transfer date back to the Ampere's and Faraday's laws which are part of the well-known Maxwell's equations for electromagnetic fields and waves. It was the pioneer Nikola Tesla who first conceived and investigated the principle of WPT and its applications at the end of the nineteenth century. At that time, the technology associated with power electronics and micro-controllers was simply not available. Moreover, advances in magnetic materials were limited to ferromagnetic materials that were only suited to operate at low utility frequencies. Due to these limitations, WPT was widely supposed to be impractical and research works on this area were very limited. With advances in semiconductor, control and material technologies, Tesla's ideas and concepts on WPT started to find practical applications. It was not until the early 1990s when researchers at Auckland University demonstrated practical applications for rail transmit, assembly line and domestic appliances. More recently, a successful WPT demonstration was performed in 2007 by a team of researchers from the Massachusetts Institute of Technology. They claimed to be able to light a 60-W incandescent bulb located at a distance of two meters from the source without using wires. Since then, the research activity on WPT has dramatically increased over the years.

WPT has the potential to disrupt and revolutionize classical "wired" power transfer applications. Current WPT technologies have reached a commercial stage in applications going from consumer electronics (smartphones, laptops, tablets, etc.), medical implants and domestic appliances. In this book, we focus our attention on industrial applications, particularly on electrical vehicles (EVs). Current EVs sales have experienced a relevant, and in some cases remarkable, increment. Forecasts for EVs are also optimistic. WPT offer opportunities to overcome some limitations to charging operations. The recent industry trend on autonomous vehicles also fosters the interest on WPT. Researchers and car manufacturers worldwide are dedicating efforts to explore commercial applications of wireless EV charging technology. Although some basic EV wireless chargers already exist on the market, wireless charge for EVs needs more efficient functionalities. Notably, these include operation under misalignment conditions (the charger and the vehicle are not aligned) and integration with the future electrical grid, where vehicles can actively discharge their batteries to provide energy to the network. In order to implement these enhanced functionalities, a comprehensive knowledge of the fundamentals of WPT in the EV context is necessary.

This book describes the fundamentals of WPT for EV chargers. For quality purposes, the book is written from a global perspective so that all the chapters are clearly related. The structure of the book has been judiciously designed to provide easy-to-follow content. The book includes some MATLAB and Simulink files, which can help readers to better understand the particularities of the technology and even design their own EV wireless charger.

This book is intended for engineers and physicists who are interested on analysing wireless power transfer in EVs. Specifically, it will benefit those interested in gaining a comprehensive view of the design procedures for an EV wireless charger. To understand this process, the book is divided into six chapters. After an introduction to WPT technologies in Chap. 1, Chap. 2 provides a description of the particularities and requirements of EV charging. A generic architecture of wireless chargers for EVs is presented in this chapter. The architecture is based on magnetic resonant WPT, meaning that coils are an essential component of EV wireless chargers. Chapter 3 describes the different topologies used for coils in these systems. It also includes an illustrative example of how to characterise the electrical parameters of a pair of coils using a software tool. Next, Chap. 4 presents the compensation networks and explains how they can be designed for a stable operation. The coils and compensation networks are both excited with high-frequency currents. These currents are obtained from power converters. Chapter 5 provides an overview of the power converters used in EV wireless chargers. Specifically, it studies the control algorithms applied to these electronics. Once the main components of EV wireless chargers have been analysed, Chap. 6 describes a generic procedure for designing this kind of system. An illustrative example of its application is also included.

Málaga, Spain   Alicia Triviño-Cabrera
May 2019   José M. González-González
                                                                                        José A. Aguado

Acknowledgements

This book collects the know-how and experience of the authors while developing several research and industry projects at the University of Malaga, Spain. The authors are grateful to many collaborators from industry and specially graduate students who have been involved during different research and prototyping stages.

Contents

1 Fundamentals of Wireless Power Transfer 1
  1.1 Introduction 1
  1.2 Technologies 3
       1.2.1 Inductive WPT 5
       1.2.2 Magnetic Resonance WPT 7
       1.2.3 Capacitive WPT 8
       1.2.4 Strongly Coupled Magnetic Resonance WPT. 9
       1.2.5 Microwave Power Transfer. 10
       1.2.6 Optical WPT 12
       1.2.7 Summary. 13
  1.3 Associated Technologies. 13
       1.3.1 Data Communication 13
       1.3.2 Receiver Localization. 15
       1.3.3 Foreign Object Detection 16
  References 16

2 Wireless Chargers for Electric Vehicles 19
  2.1 Types of Electric Vehicles 19
  2.2 Battery Technology in EVs. 20
       2.2.1 Battery Management System 23
  2.3 Charging Modes in EVs 25
  2.4 Benefits of WPT. 27
  2.5 WPT Operation Modes. 29
  2.6 Vehicle to Grid Technology 31
  2.7 Standards for EV Wireless Chargers 31
       2.7.1 SAE J2954 32
       2.7.2 IEC 61980 33
       2.7.3 ISO 19363 33

2.8   Wireless Power Transfer Market Perspectives.................   34
      2.8.1   WiTricity   34
      2.8.2   Qualcomm..   35
      2.8.3   EVATRAN   35
      2.8.4   Eaton HEVO..   36
      2.8.5   Momentum Dynamics   36
      2.8.6   Car Manufacturer Wireless Chargers..................   36
2.9   Generic Structure of the EV Wireless Charger   37
References   39

3   Coil Design for Magnetic Resonance Chargers   43
    3.1   Introduction   43
    3.2   Coil Geometry and Materials............................   45
    3.3   Electromagnetic Emissions of Two Coupled Coils   52
    3.4   Use of Ferromagnetic Material   55
    3.5   Magnetic Shielding.....................................   58
          3.5.1   Conductive Shielding...........................   58
          3.5.2   Active Shielding   63
          3.5.3   Reactive Resonant Shielding   65
    References   65

4   Compensation Networks   69
    4.1   Introduction   69
    4.2   Stability of the Compensation Networks   72
    4.3   Mono-resonant Compensation Networks   72
          4.3.1   Series–Series Topology   73
          4.3.2   Series–Parallel Topology   77
          4.3.3   Parallel–Series Topology   79
          4.3.4   Parallel–Parallel Topology   84
          4.3.5   Discussion.....................................   87
    4.4   Multi-resonant Compensation Networks...................   89
          4.4.1   LCL Topology   90
          4.4.2   LCC Topology   92
    References   98

5   Power Electronics   101
    5.1   Power Electronics......................................   101
    5.2   Semiconductor Devices for Power Converters   103
    5.3   Uni-directional Charger.................................   106
          5.3.1   Grid Rectifier   108
          5.3.2   Primary Inverter................................   113
          5.3.3   Secondary Rectifier   118
          5.3.4   Secondary DC/DC Converter   119

	5.4	Bi-directional Charger	120
		5.4.1 Primary AC/DC Converter	120
		5.4.2 Secondary AC/DC Converter	124
		5.4.3 Secondary DC/DC Converter	124
	References	125

6 Design Procedure of an EV Magnetic Resonance Charger 129
	6.1	Introduction	129
	6.2	Illustrative Design Procedure	135
		6.2.1 Requirement Specifications	135
		6.2.2 Selection of Potential Configurations	136
		6.2.3 Design of the Coils and Compensation Networks	136
		6.2.4 Design of the Power Converters	137
		6.2.5 Control Design	138
		6.2.6 Software Tests	140
	6.3	Prototype Implementation	141
		6.3.1 The Coupler	142
		6.3.2 Compensation Networks	143
		6.3.3 Power Converters	143
		6.3.4 Controllers	145
		6.3.5 Load Modelling	147
		6.3.6 Summary of the Main Parameters of the Implemented Prototype	147
	6.4	Lab Tests	148
	References	151

Appendix: Software Tools 153

Index 161

Acronyms

AC	Alternating Current
AWG	American Wire Gauge
BEV	Battery Electric Vehicle
BJT	Bipolar Junction Transistor
BMS	Battery Management System
BWPT	Bi-directional Wireless Power Transfer
CC	Constant Current
CC-CV	Constant Current-Constant Voltage
CET	Contact-less Energy Transfer
CFR	Code of Federal Regulations
CLCL	Capacitor-Inductor-Capacitor-Inductor
CV	Constant Voltage
DC	Direct Current
EV	Electric Vehicle
FET	Field Effect Transistor
FOD	Foreign Object Detection
GTO	Gate Turn-off Thyristor
HEV	Hybrid Electric Vehicle
ICE	Internal Combustion Engine
ICNIRP	International Commissioning of Nonionizing Radiation Protection
IEC	International Electrotechnical Commission
IEEE	Institute of Electrical and Electronics Engineers
IGBT	Insulated Gate Bipolar Transistor
ISO	International Organization for Standardization
LCC	Inductor-Capacitor-Capacitor
LCL	Inductor-Capacitor-Inductor
LCO	Lithium-Cobalt Oxide
LFP	Lithium Iron Phosphate
MCC	Multi-stage constant-current
MOSFET	Metal Oxide Semiconductor Field Effect Transistor

MPT	Microwave Power Transfer
NCA	Lithium Nickel Cobalt Aluminium Oxide
NFC	Near Field Communication
NiMH	Nickel Metal Hydride
NMC	Lithium Nickel Manganese Cobalt Oxide
OLEV	On-line Electric Vehicle
PFC	Power Factor Corrector
PHEV	Plug-in Hybrid Electric Vehicle
PLL	Phase Locked Loop
PP	Parallel-Parallel
PS	Parallel-Series
PWM	Pulse Width Modulation
RF	Radio Frequency
RFID	Radio Frequency Identification
RPEV	Roadway Powered Electric Vehicle
SAE	Society of Automotive Engineers
SAR	Specific Absorption Rate
SEPIC	Single-Ended Primary Inductor Converter
SHARP	Stationary High Altitude Relay Platform
SiC	Silicon Carbide
SoC	State of Charge
SoH	State of Health
SP	Series-Parallel
SS	Series-Series
SSPS	Space Solar Power Satellite
THD	Total Harmonic Distortion
UAV	Unmanned Aerial Vehicle
V2G	Vehicle-to-Grid
VA	Vehicle Assembly
WPT	Wireless Power Transfer
ZCD	Zero Crossing Detector
ZEBRA	Zero Emissions Batteries Research Activity (Sodium Nickel Chloride battery)
ZPA	Zero Phase Angle

Symbols

B	Magnetic flux density (T).
C_1	Capacity of the primary compensation network capacitor in uni-resonant systems. For multi-resonant systems, capacity of the primary compensation network series capacitor (F).
C_2	Capacity of the secondary compensation network capacitor in uni-resonant systems. For multi-resonant systems, capacity of the secondary compensation network series capacitor (F).
C_{f1}	Capacity of the parallel capacitor of the primary compensation network in multi-resonant systems (F).
C_{f2}	Capacity of the parallel capacitor of the secondary compensation network in multi-resonant systems (F).
D	Duty cycle.
D_{avg}	Average diameter.
d_{in}	Internal diameter.
d_{out}	External diameter.
E	Electric field (kV/m).
η	Efficiency.
f	Frequency (Hz).
H	Magnetic field strength (A/m). Also used as magnetic field in the presented of a shield.
H^i	Magnetic field in the absence of a shield.
I_1	Current flowing through the primary coil (A).
I_2	Current flowing through the secondary coil (A).
I_{C_1}	Current flowing through C_1 (A).
I_{C_2}	Current flowing through C_2 (A).
I_{in}	Input current of the analysed system (A). Used in the analysis of power electronics.
I_L	Current flowing through the battery or its equivalent resistance (A).
I_{out}	Output current of the analysed system (A). Used in the analysis of power electronics.

I_s	Source current of the analysed system (A). Used in the analysis of compensation networks.
L	Self-inductance (H).
L_1	Self-inductance of primary coil (H).
L_2	Self-inductance of secondary coil (H).
L_{f1}	Self-inductance of multi-resonant primary compensation coil (H).
L_{f2}	Self-inductance of multi-resonant secondary compensation coil (H).
M	Mutual inductance (H).
N	Number of turns.
ω	Frequency (rad/s).
ω_0	Operational frequency (rad/s).
P_{loss}	Power losses (W).
q_1	Primary side quality factor.
q_2	Secondary side quality factor.
R_1	Resistance of primary coil (Ω).
R_2	Resistance of secondary coil (Ω).
R_{eq}	AC equivalent resistance (Ω). Used to analyse a circuit when a full-bridge rectifier is included.
R_L	Equivalent resistance to the battery (Ω).
SE_H	Effectiveness of shielding.
μ_0	Vacuum permeability (H/m).
V_1	Voltage measured before the compensation system on the primary side (V).
V_2	Voltage measured before the compensation system on the secondary side (V).
V_{Grid}	Grid voltage (V).
V_{DC}	Direct Current voltage (V).
V_{in}	Input voltage of the analysed system. Used in the analysis of power electronics (V).
V_L	Voltage measured in the battery terminals (V).
V_{out}	Output voltage of the analysed system. Used in the analysis of power electronics (V).
V_s	Source voltage of the analysed system. Used in the analysis of compensation networks (V).
VAr_1	Reactive power consumed on the primary side (VAr).
VAr_2	Reactive power consumed on the secondary side (VAr).

Chapter 1
Fundamentals of Wireless Power Transfer

1.1 Introduction

Wireless Power Transfer (WPT) is the technology by which one or multiple transmitters generate an electromagnetic wave, which is processed by one or several receivers without any type of conductor in order to extract power from the wave. In contrast to wireless communication systems, the electromagnetic wave in WPT systems is used by the receiver to store energy in a battery or to power electronics.

The first experiments on wireless power transfer were performed by the engineer Nikola Tesla at the end of the 19th century. As described in [19], he was able to transmit power with microwaves between two objects 48 km apart. Another of Tesla's experiments consisted in powering 200 bulbs without cables, from a power source located 25 miles away. For these experiments, issues related to human and electrical safety were not considered.

It was not until the 21st century that the research community regained an interest in WPT systems. This renewed motivation was driven by the development of power converters in that period, which allowed the use of frequency in the range of dozens of kHz and kW operations. This had not been possible previously.

In this new trend, the technology was initially referred to as Contactless Energy Transfer (CET). However, "wireless power transfer" ultimately became the accepted term.

WPT technology is now a reality. We find this technology supported in commercial products such as electric toothbrushes, power mats for mobile phones and even chargers for electric vehicles (EVs). In 2017, 450 million units incorporating this capability were sold globally, primarily in smartphones, smartwatches and small home appliances. This figure represented a 75% increase on sales recorded the previous year. This significant increase is expected to continue in the near future. In fact, IHS Markit predicts that this market sector will grow to more than 2.2 billion units by 2023 [3]. This expansion will also have significant economic benefits: Navigant Research estimates that the revenue from wireless chargers will be close to 17.9 billion dollars by 2024 [2].

© Springer Nature Switzerland AG 2020
A. Triviño-Cabrera et al., *Wireless Power Transfer for Electric Vehicles: Foundations and Design Approach*, Power Systems,
https://doi.org/10.1007/978-3-030-26706-3_1

If we focus more closely on WPT applications, we can observe that they are implemented in diverse ways. Consequently, WPT systems can be classified according to the following criteria (please refer to Fig. 1.1):

- **Transferred power**. WPT systems comprise applications for transmitting low power (up to 1 kW), medium power (1–100 kW) and high power (more than 100 kW). The power requirement of the application greatly impacts on the system design. Thus, for low power applications, efficiency is not as crucial as in other kinds of systems. Instead, transferring the maximum power possible is usually the primary aim of low-power applications.
- **Uni-directional or bi-directional** power transfer. According to this criterion, we can differentiate between WPT systems where the power transfer is always originated by a fixed element where a source is connected. This scheme corresponds to a uni-directional WPT. Alternatively, there are bi-directional systems where the load (a battery or a capacitor) occasionally provides energy to the source.
- **Gap**. This term refers to the distance between the energy transmitter and the receiver. Although all WPT systems avoid cables between these two components, in some applications there must be a contact between them. This is the case with power mats. Alternatively, in some applications the transmitter and the receiver are separated by several centimeters or even meters.
- **Capacity to operate with intermediate objects** in the gap between the power transmitter and the receiver. Due to the wavelength, some technologies cannot operate with intermediate objects, others suffer from a relevant degradation under the presence of these elements, whereas in other technologies the impact is not noticeable.
- **Number of transmitters**. The most simple topology for a WPT system consists of one power transmitter and one power receiver. In order to extend the WPT spatial operability, several transmitters can be deployed in a region in order to transfer power to a load. In this case, more than one transmitter can be activated simultaneously considering their power availability and the efficiency of the power transfer (e.g. their power resources when derived from renewable energy sources). On the other hand, the role of transferring power can be executed by a different transmitter in a different time interval. This could be appropriate for mobile loads.
- **Number of receivers**. Although the usual topology for WPT systems considers just one receiver, there are some configurations designed to support multiple loads. Thus, it is possible that multiple receivers can benefit from the power generated by one transmitter.
- **Stationary/Mobile receiver**. In some applications, WPT must be able to handle the receiver being placed in a random position before the charge starts. This is the case for dynamic EV wireless charging.
- **Medium**. Although most current WPT products operate with an air gap between the power transmitter and the receiver, this technology can also be applied in other mediums such as water [25], ground [14] or biological tissue [13]. The medium clearly impacts on the efficiency as it is responsible for the power transmission losses. For instance, the study carried out in [25] examined how the efficiency of the underwater WPT system is up to 5% lower than an air-gap system.

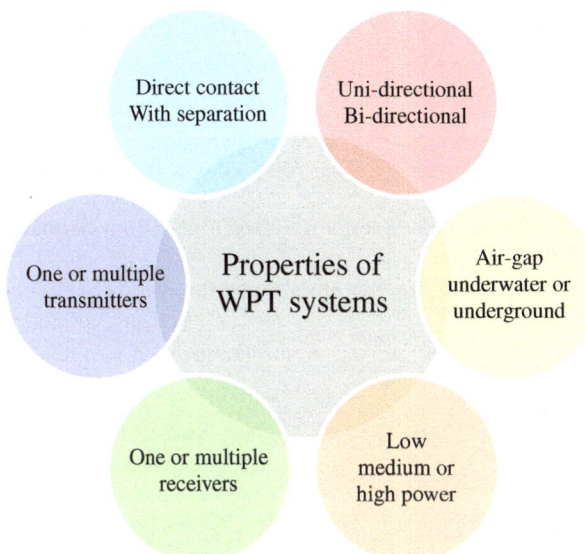

Fig. 1.1 Main features of WPT systems

1.2 Technologies

In all these previous experiments and the ensuing work, wireless power transfer is supported by an electromagnetic wave travelling from the power emitter to the power receiver. In WPT systems, the electromagnetic field is exclusively generated to transfer power. Conversely, energy harvesting techniques make use of the electromagnetic waves generated to transfer information to acquire energy to power devices. Thus, energy harvesting techniques are restricted to the requirements imposed by the information transfer, which are not present in WPT technologies.

Figure 1.2 illustrates the generic diagram of a WPT system. The maximum dimension of the power emitter (the antenna) is L_{DEV}. The transmitter and the receiver are separated a distance d, usually referred to as the gap. Electromagnetic waves are characterized by their wavelength λ or their frequency f.

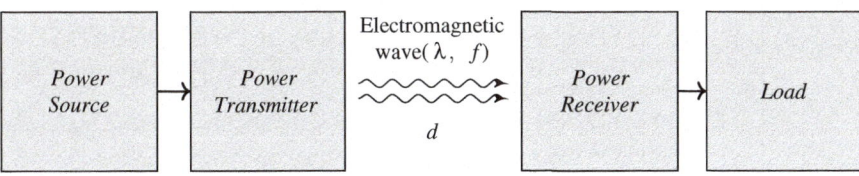

Fig. 1.2 Generic diagram of a WPT system

The behaviour of an electromagnetic wave is defined by Maxwell's equations. These complex equations can be simplified when some conditions hold, leading to the near-field and far-field operation. Both scenarios are described next.

- **Near-field operation or non-radiative propagation**. Three conditions must be satisfied to work in this kind of scenario. They are:

1. The size of the transmitter element, referred to as L_{DEV}, is much smaller than the wavelength λ.
2. The distance between the energy emitter and the receiver is much smaller than the wavelength λ.
3. The distance between the transmitter and the receiver is much smaller than $2 \cdot (L_{DEV}^2)/\lambda$.

- **Far-field operation or radiative propagation**. This is based on the electric field of the electromagnetic wave. In this case, the conditions are:

1. The distance between the energy emitter and the receiver is greater than the wavelength λ.
2. The size of the transmitter element L_{DEV} is more than 10 times greater than the wavelength λ.

In each scenario, there is a group of WPT technologies as presented in the chart below. Thus, in the near-field operation we have the inductive, the resonant and the capacitive wireless power transfer. Alternative, Microwave-based or optical WPT are far-field technologies. There is an intermediate configuration, referred to as Strongly Coupled Magnetic Resonance systems, which belongs to the intermediate operation between the near-field and the far-field technologies. It is important to know which group a WPT technology belongs in, so as to analyse the electrical systems correctly. Specifically, Maxwell's equation can be simplified with Kirchhoff's Law in the near-field operation while RF analysis and optics-based equations are necessary for the operation in the far-field.

All the aforementioned WPT technologies are described next (Fig. 1.3).

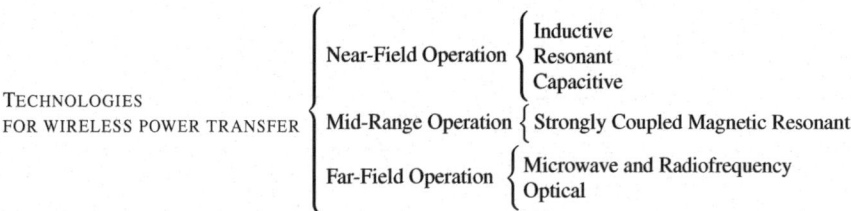

Fig. 1.3 Classification of WPT technologies

Fig. 1.4 Illustration of
induced voltage due to
varying magnetic field

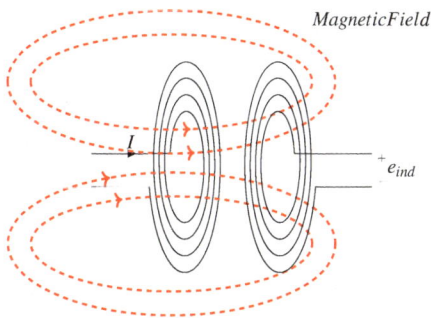

1.2.1 Inductive WPT

Inductive WPT is realized with the magnetic field of the electromagnetic wave. The
operation principle is explained by the interaction of the magnetic and electrical
behaviour described by Ampère's Law and Faraday's Law.

According to Ampère's Law, a current-carrying wire generates a magnetic field
around it. The intensity of the magnetic field and its orientation depend on the topology of the wire. Specifically, Ampère's Law states that:

$$\oint \overline{H} dl = I \tag{1.1}$$

where $\overline{H}$ is the magnetic field intensity of the magnetic field generated by the electric
current I and dl is the differential element of length along the path on which the
current travels. As a consequence of this physical phenomenon, the frequency at
which the intensity of the magnetic field varies is equal to the frequency of the current
in the wire. Figure 1.4 illustrates the magnetic field of some common structures
employed in inductive wireless chargers.

As shown, coils are able to concentrate the magnetic field around the area in which
they are defined to a higher degree than a simple wire.

As described by Ampère's Law, when a time-varying current passes through a
coil, a time-varying current magnetic field is generated around this element. If that
time-varying magnetic field traverses a different coil, a voltage (e_{ind}) is induced in
its terminals. This effect is described by Faraday's Law as follows:

$$e_{ind} = -\frac{d\phi}{dt} \tag{1.2}$$

where ϕ is the flux of the magnetic field passing in the area limited by the coil.

The combination of these two phenomena forms the basis of the inductive and
other magnetic-based WPT technologies. Inductive WPT technology requires a pair
of coils referred to as the primary and secondary coils. This is presented as a diagram in Fig. 1.5. In the primary coil, a time-varying current I_S must be produced

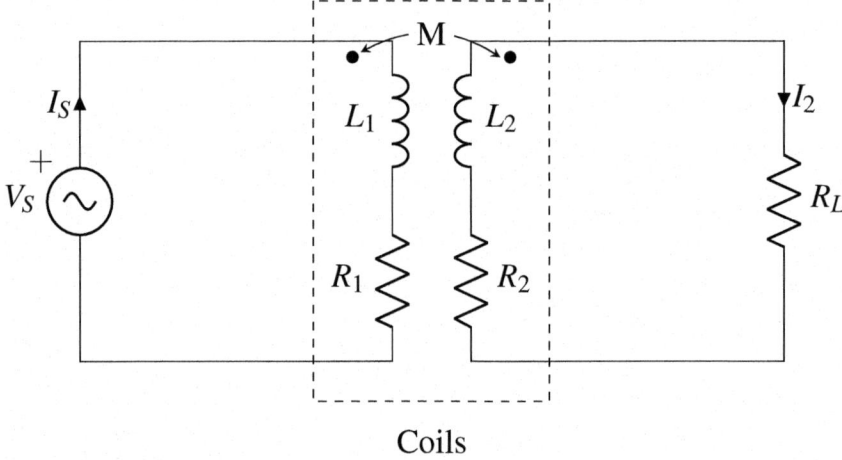

Coils

Fig. 1.5 Equivalent circuit of inductive WPT

by a generator. The magnetic field resulting from this must traverse the area of the secondary coil to which the load to be powered/charged (R_L) is connected. Between the generator and the primary coil, there are usually intermediate electronic components. Similarly, there are other electric systems between the secondary coil and the load. These additional elements are included to improve the wireless power transfer efficiency as explained next.

In general terms, we can state that the best approach is to produce an induced voltage that is as high as possible. As shown by Faraday's Law, the induced voltage is proportional to the rate of change of the flux traversing the secondary coil. This means that a coil traversed by two magnetic fields with the same magnitude but different frequencies at two distinct moments will experience two different induced voltages. When the magnetic field passing through the coil is of the highest frequency, it will result in a higher induced voltage. Thus, the variation of magnetic flux in the secondary coil should preferably be as high as possible.

Thus, it is of interest for an inductive-based WPT to hold these two conditions:

- Most of the magnetic field generated by the primary coil traverses the secondary coil.
- The frequency of the magnetic field involved in the WPT is as high as possible while allowing for a near-field operation.

The first condition initially implies that big coils are preferable on the secondary side, but the application imposes some limits for this component in terms of size, weight and cost. This restriction is clearly observed in biomedical applications. With regard to EV applications, there is a limit to the size of the coils because of the structures in which the WPT components must be inserted and the cost of the materials. Please note that WPT for EVs is not supported by inductive WPT but by advanced technologies based on this kind of magnetic WPT.

Considering that inductive WPT also benefits from a higher rate of flux change, the main strategy for enhancing the WPT in inductive systems is to increase the frequency of the electrical current in the primary coil. This will lead to an increase in the frequency of the magnetic field and, consequently, the rate of flux change is also increased. Power converters are part of the magnetic-based WPT systems in order to elevate the operational frequency.

Radio Frequency Identification (RFID) and Qi are commercial technologies that are based on inductive WPT.

1.2.2 Magnetic Resonance WPT

Magnetic resonance or resonant WPT can be considered an improvement on inductive WPT in which the electrical system is forced to work under resonant conditions. To meet this requirement, the pair of coils is connected to structures composed of reactive elements such as capacitors or additional coils. These structures are referred to as the compensation networks. Figure 1.6 shows the generic diagram of a resonant WPT system.

The most simple compensation topologies consist of a single capacitor, which may be connected to the primary and the secondary in series or in parallel. These networks are referred to as mono-resonant compensation topologies.

Alternative, more complex compensation topologies are also an option. These are identified as multi-resonant compensation topologies.

Wireless charge in EVs is based on resonant technology. Thus, more details about the compensation topologies and other relevant aspects that ensure operation under resonant conditions are described in the following chapters.

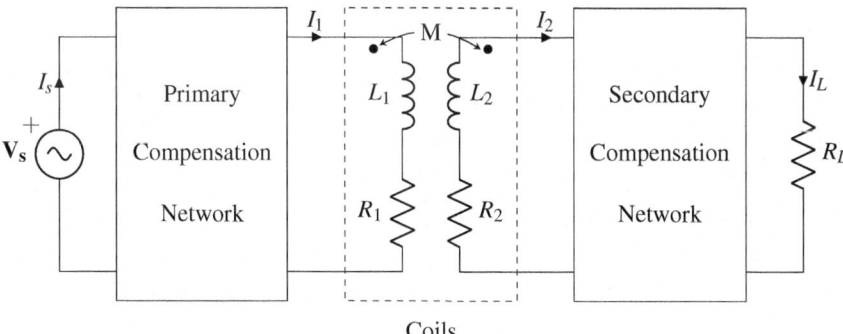

Fig. 1.6 Generic diagram for magnetic resonance wireless chargers with compensation networks

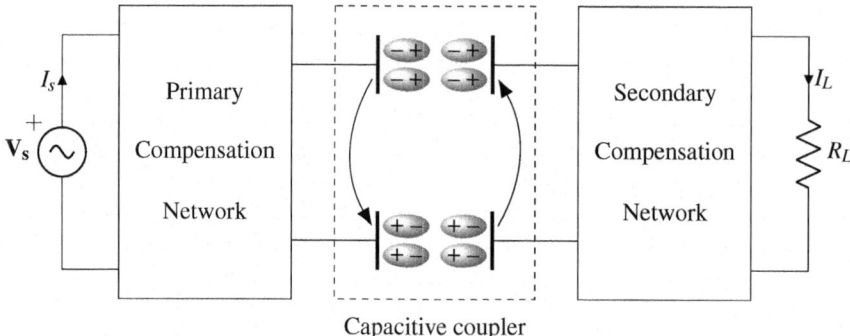

Fig. 1.7 Generic diagram for capacitive wireless chargers with compensation networks

1.2.3 Capacitive WPT

In contrast to the previous near-field WPT techniques, capacitive WPT is achieved by means of the electrical field. To work with a capacitive WPT, two metallic plates are inserted in the power emitter and the receiver. In each part, one plate is connected to each end of the conductors connecting the power source or the load as shown in Fig. 1.7.

The plates in each end of the emitter are parallel with the corresponding plates in the receiver. When the two pairs of plates are close enough, they act as two capacitors and, consequently, the electrical circuit is closed. The capacitors are identified as the forward and the return capacitors.

Under the circumstances described above, capacitive coupling takes place. Thus, an electric field is generated between the plates and, as a result, an electrical current is induced in the power receiver. Similarly to the magnetic-based WPT, the current induced in the power receiver is proportional to the rate of change of the electric field flux between the two pairs of plates. In order to increase this rate, the operational frequency is elevated from the power provided by the grid. Power converters are utilized for this purpose so that V_s is generated.

In contrast to magnetic-based techniques, capacitive WPT is able to transfer even with metallic objects [15]. The presence of these intermediate metallic objects does not greatly impact on the process as the losses are not relevant and the objects themselves do not even reach an excessive temperature in applications where capacitive WPT has been tested.

Another significant advantage lies in the fact that the electrical field is restricted to the region separating the two plates. As a result, the field of interest does not escape from this area, although this did occur with the magnetic field in the WPT techniques previously studied. At this stage, resonant technology seems to be developed enough to support a higher amount of power transfer [8].

1.2.4 Strongly Coupled Magnetic Resonance WPT

Strongly Coupled Magnetic Resonance (SCMR) or WPT based on magnetically coupled resonance are terms used indistinctly in the related literature.

In contrast to the previous technologies, SCMR systems are considered mid-range WPT as they do not fit perfectly in the near-field or the far-field operation. Although $L_{DEV} << \lambda$, the other two conditions put forward for near-field operation do not hold. In particular, on mid-range operations, it is also verified that $1 < d < 10 * L_{DEV}$ but the L_{DEV} is up to several times the distance separating the power source and the load.

The first relevant experiment involving SCMR technology was carried out by a group of researchers at the Massachusetts Institute of Technology in 2007. In this experiment, they successfully powered a light bulb without cables, with the power source located 2 m away from this component. The efficiency of the power transfer in this case was around 40%.

This experiment established the main properties of SCMR circuits. They are:

- **4-coil topology**. The most common configuration for SCMR circuits is a topology composed of four coupled coils: two in the power source (the driver and the transmitter) and two in the load (the receiver and the load). None of them is connected with conductors; instead they are magnetically coupled. The topology is illustrated in Fig. 1.8. This Figure also indicates the coupling coefficient between the coils.
- **The intermediate coils are self-resonant**. The transmitter and the receiver coils are used as self-resonant structures, meaning that the operational frequency of the system is set according to the free resonant frequency of these coils. This parameter is computed based on the parasitic capacitance of the intermediate coils. Taking into account the physical properties of the coils, the free resonant frequency is usually set in the interval from 100 kHz to 20 MHz.
- **Compensation topologies in the driver and the load loops**. The coils connected to the power source and to the load are a simple loop. As their parasitic capacitance is not relevant, it is necessary to incorporate lumped reactive components to ensure that the system works under resonant conditions.

Fig. 1.8 4-coil topology in SCMR systems

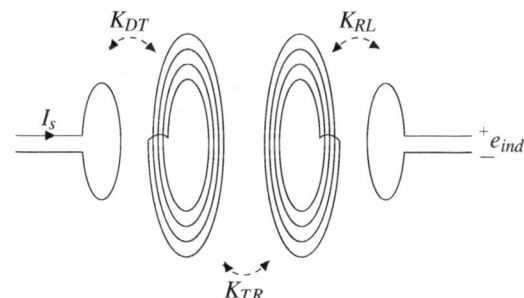

The capacity of SCMR systems to function in the mid-range scenario depends on the 4-coil or indirect-link topology. To understand its properties, we can rely on an equivalent two-coupled coil system. In this model, the coupling coefficient is simplified as $K_{DT} * K_{RL}/K_{TR}$, where K_{DT} is the coupling coefficient between the driver and the transmitter, K_{RL} is the coupling coefficient between the receiver and the load and K_{TR} is the coupling coefficient between the transmitter and the receiver. The coupling coefficient indicates how much of the magnetic flux of a coil is used by the other coupled coil to induce a voltage. For an efficient wireless power transfer, it is preferable to obtain coupling coefficients close to 1 while satisfying other system requirements such as the distance between the power source and the load. In general terms, the more separated two coupled coils are, the lower the coupling coefficient becomes.

If we focus on the equivalent coupling coefficient, we can observe that this parameter can be close to 1 even when the coupling coefficient between the transmitter and the receiver coils is low. This can be achieved by properly adjusting the other two coupling coefficients K_{DT} and K_{RL}. In fact, some control algorithms for SCMR systems dynamically alter the distance between the driver loop and the transmitter coil and, as a consequence, the coupling coefficient K_{DT}.

As explained in [22], control algorithms in SCMR systems are designed with two main goals in mind: obtaining the maximum power transfer in the load or maximizing the efficiency of this process. To achieve one of these goals, the system adjusts some parameters. In particular, the strategies are (i) modifying the operational frequency; (ii) including adjustable matching networks; and (iii) altering the distance between the driver and the transmitter coils, as previously mentioned. For these three approaches, it is necessary to consider the phenomena of frequency splitting and frequency bifurcation. Frequency bifurcation commonly occurs in resonant structures and will be described in detail in Chap. 4. For more information about the frequency splitting phenomenon, please refer to [22].

1.2.5 Microwave Power Transfer

Microwave Power Transfer (MPT) refers to WPT based on microwave to transfer energy in a far-field context. The procedure can also be extended to Radio-Frequency (RF) signals with minor modifications. In fact, sometimes MPT and RF power transmission are grouped into the wave-based WPT.

The generic structure of a MPT is depicted in Fig. 1.9.

From a high-voltage DC generator, a magnetron creates a microwave signal. The microwave is then sent through the antenna. The receiver processes the signal by means of a rectenna to convert the microwave signal to a DC signal. Finally, the DC signal is received by the power electronics device.

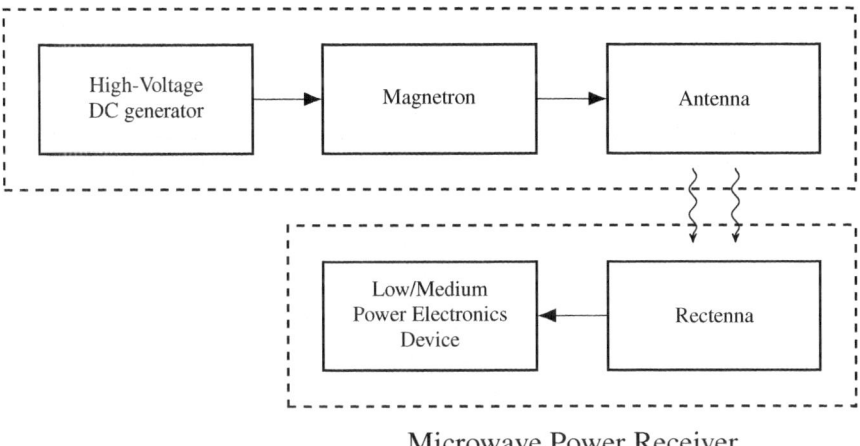

Fig. 1.9 Generic diagram of a MPT system

As can be observed, the main elements of the MPT are:

- Magnetron. This is a vacuum tube that acts as an oscillator. From a DC excitation, the magnetron generates a microwave signal without any amplification. The frequency of the generated signal is related to the dimensions of the magnetron.
- Antenna. This corresponds to the electrical component that is able to generate an electromagnetic wave to travel to the receiver. The simplest antennas generate the signal in an isotropic way, that is, the signal has the same power in all the spatial directions. Alternatively, beamforming is possible with an array of antennas controlling the difference in the current phase [16].
- Rectenna. This stands for an antenna and a rectifying circuit built in the same component. It was designed by Charles Brown in 1960 to operate with a 2.45-GHz signal.

The first relevant experiment on MPT was conducted in 1964. The group led by William Brown powered a model aircraft with a microwave power transmission. In 1968 Peter Glaser introduced the concept of Space Solar Power Satellite (SSPS). As explained in [20], this is a current research guideline in which a satellite reflects the energy from the Sun to transfer energy to the Earth. Based on MPT, in 1975 they also succeeded in transferring 450 kW to a receiver placed 1 mile away from the transmitter. In the 1980s, the Stationary High Altitude Relay Platform (SHARP) project fueled small aeroplanes flying at an altitude of 21 km. From this altitude, the aeroplanes provided telecommunication services. A 500-kW power transfer was achieved with a 5.8-GHz signal. The group led by Professor Shinohara from Kyoto University has been developing microwave-based wireless chargers for EVs, achieving up to 84% efficiency in some of their prototypes [12].

MPT is currently used in home environments in some commercial implementations. One of these is Powercast [6]. Powercast source generates a 915-MHz signal to transfer up to 3 W with a directional beam or up to 1 W for an isotropic transmission. Cota, the product developed by Ossia for WPT, is also supported by RF transmission [4]. It implements advanced algorithms to detect obstacles or human beings in the transmission path. A Bluetooth channel is also required for its operation. Energous has developed WattUp for WPT in the range of 2.4 and 5 GHz [1]. The key feature of WattUp is its use of an array of antennas in the transmitter to control the beamforming and make the power transfer more efficient. The receiver also has multiple antennas to capture more energy from different electromagnetic waves. The gap can be up to 5 m with a maximum efficiency of 70%. It can be observed that the commercial applications are low-power systems. MPT becomes unsafe for higher levels of power.

1.2.6   Optical WPT

Optical WPT systems rely on electromagnetic waves with a frequency in the THz range for the power transfer. This type of wave requires the power transmitter and its receiver to be in the line-of-sight; that is, without any intermediate obstacles, as the wave cannot traverse them. If this condition holds, the gap can be up to several kilometres. This was the case with the first experiments on optical WPT. The first experiment with this technology consisted in powering a Mini Rover over a distance up to 280 m. The output power of the laser was 5 W whereas the power required by the vehicle was 1 W [11]. Three years later, in 2003, another experiment was conducted by NASA. With a 1-kW laser beam, they were able to remotely power a 6-W drone [9]. Other experiments with UAVs have been reported in the literature but the power and efficiency of the transfer were still reduced [11]. Moreover, this directivity can also bring an important benefit in that a controlled optical WPT does not interfere with other systems. However, nowadays electronics cannot easily adjust the direction of the beam, but this can be performed with passive elements (e.g. lens). Another clear disadvantage is the poor efficiency of these systems (around 25%) [27]. The progress on laser technology and photovoltaic cells could improve this metric in the near future.

A generic diagram of an optical WPT is shown in Fig. 1.10. According to this configuration, the laser diode in the transmitter is controlled by the current mode in the generator. A beam director serves to adjust the direction of the power transfer. In the receiver, a photovoltaic cell converts the received light into power with the corresponding power converters. The DC power is then used to power a load or a battery.

For low-power applications, optical WPT systems (specially in the infrared region) are safe to be used in human environments. This has prompted the commercialisation of optical WPT products for homes. Wi-Charge is an illustrative example of infrared optical WPT systems for low-power devices [5]. In its conventional configuration,

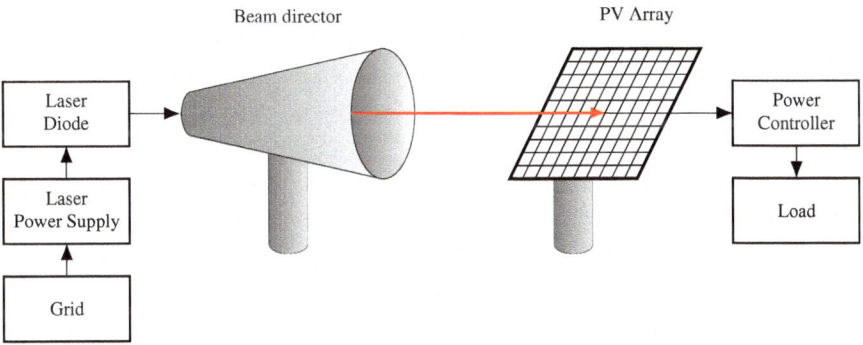

Fig. 1.10 Generic diagram of an optical WPT system

the transmitter is usually placed on the ceiling of a room. It is equipped with advanced electronics that can modify the beam so it can be aligned with the receiver. According to the specifications, they can charge up to 3 W in 1 km.

1.2.7 Summary

WPT comprises multiple technologies. The previous sections have explained the fundamentals of the near-field, mid-range and far-field technologies. Table 1.1 summarises the main advantages and disadvantages of these technologies.

1.3 Associated Technologies

Real WPT products are complex systems that involve not only technology for transferring power through the medium but also additional electronics and mechanisms to ensure that this operation is performed in a controlled way. The controls put in place seek a safe and feasible operation in terms of electrical and magnetic performance. In addition, the system has to be integrated with other systems or infrastructures.

The following subsections describe the most relevant technologies that are included in real commercial WPT products.

1.3.1 Data Communication

Devices and their batteries have to be charged in a specific way ensuring that some electrical magnitudes are maintained in a given interval. For instance, some batteries

Table 1.1 Properties of WPT technologies

WPT technologies		Key advantages	Potential disadvantages
Near field	Capacitive	• Medium power transfer up to several kilowatts • Transfer even with metal objects between the transmitter and the receiver • It relies on metal plates to transfer power, which are cheap components • Suitable for small size applications with a reduced gap (up to 10 cm) • Electric field restricted to the area between the plates: no need for controlling EMI	• Efficiency at the range of 70–80% • The amount of power transfer is highly dependent of the gap Reduced gap (up to 10 cm)
	Inductive	• Simple implementation for low power applications • Galvanic isolation provided • Simple control	• Limited efficiency at the range of 20% • Short transmission gap (up to several cms) • Need for controlling the electromagnetic emissions • Highly dependent on the presence of obstacles (especially the metallic ones) in the area between the transmitter and receiver and its surroundings
	Resonant	• High power transfer up to several kilowatts • Mature technology with commercial applications already • Power transfer is enabled even with some variations on the receiver's position (although the efficiency decreases) • Galvanic isolation provided	• Expensive components for high power applications • Highly dependent on the presence of obstacles (especially the metallic ones) in the area between the transmitter and receiver and its surroundings • Need for controlling the electromagnetic emissions
Mid-range	SCMR	• Long effective transmission distance up to 2–3 m • Suitable for mobile receivers • Suitable for low and medium power applications	• Efficiency lower than 50% • Complex implementation of the coils and the control algorithms • Bulky components in the receiver and the transmitter
Far field	Microwave	• Long effective gap up to several km • It allows dynamic configuration of the microwave beam to adapt the power transfer with mobile loads • Potential to transfer several kilowatts power	• Low efficiency less than 10% for high power applications with long gaps • Complex implementation • Unsafe operation with living beings between the power transmitter and the receiver • Considerable receiver's dimension
	Optical	• Long effective gap up to several kilometers • It allows dynamic configuration of the light beam to adapt the power transfer with mobile loads • Potential to transfer several kilowatts power • Reduced size of the transmitter	• Low efficiency (around 20%) • No obstacles allowed: there must be a line-of-sight from the transmitter to the receiver • Unsafe operation with living beings between the power transmitter and the receiver if the infrared region is not used

in EVs require a constant current during part of the charging, and after that the voltage is kept fixed. In order to comply with these restrictions, the components in the wireless transmitter must adjust the method by which the power signal is generated according to the status of the battery. Some research works have proposed methods to infer the status of the battery from measurements taken in the transmitter but these are very sensitive to the tolerance of the components [7, 21, 24]. Thus, it is necessary to transfer information about the battery to the power transmitter. Moreover, the wireless power transfer can be controlled not only electrically but following an operational restriction. In that regard, some public infrastructures prevent WPT from being initiated until the user/device is authenticated. Alternatively, private chargers demand user information for billing purposes. During the charge, energy prices may also fluctuate, making it necessary to inform the power receiver about these economic parameters. These procedures are supported by data exchange between the power transmitter and the receiver. In wireless chargers, this communication is logically implemented through a wireless channel in order to continue the power transfer without a cable.

Thus, advanced wireless chargers incorporate established wireless systems based on cellular or ad-hoc communications (IEEE 802.11 or IEEE 802.16). In magnetic-resonance chargers, some experimental prototypes suppress the need for an additional communication system by also using the power coils to transmit information [23]. Security of wireless data transfer for wireless chargers has also been analysed in recent publications [26]. In particular, these have focused on how to make WPT unavailable for unauthorised users. In their proposal for a magnetic-resonance charger, this is achieved by varying the frequency and the duration of the pulses. Then, the capacitor of the compensation network in the receiver is adjusted.

1.3.2 Receiver Localization

Some WPT systems allow the receivers to be mobile. In order to get an efficient performance from these chargers, the transmitter needs to generate the electromagnetic wave according to the instantaneous position of the receiver. This is done differently depending on the WPT technique. Far-field WPT techniques opt for beamforming in order to transmit the power to the target destination. To acquire this capability, some transmitters are designed as an array of sources. The beam is then controlled with the difference in the phase of the currents generated by the sources.

Alternatively, near-field WPT techniques adjust the levels of the generated waves to transmit only the necessary power to the receiver. In addition, multiple transmitters may be available to be activated when needed. This is a common implementation in power mats or in dynamic EV wireless chargers. In these structures, it is necessary to locate the receiver to turn on only those transmitters that could make an effective power transmission. As explained in [17], the mutual inductance can be used as a strong indicator of the proximity between the receiver and a transmitter coil for magnetic power transfer.

1.3.3 Foreign Object Detection

The performance of some WPT technologies may be sensitive to the presence of intermediate objects between the power transmitter and the receiver. For RF-based WPT, these objects are almost irrelevant, especially with omnidirectional transmission. The opposite behaviour is observed with optic WPT, in which these items interrupt the power transmission. Magnetic-based WPT could continue the power transfer but its efficiency is greatly affected as more magnetic flux does not traverse the receiver coil.

In addition to their impact on efficiency, intermediate objects can cause unexpected accidents. For EV wireless chargers, the magnetic field involved in the power transfer induces eddy currents in the objects. As a result, the temperature of the object increases, which could eventually lead to combustion.

These effects must be taken very seriously if living beings are involved. For this reason, Wi-Charge interrupts the power transfer when an intermediate object/being is detected [5].

The Foreign Object Detection (FOD) algorithms aim to identify situations in which these objects/beings are in the intermediate zone between the transmitter and the receiver in order for the power transfer to be interrupted. In magnetic-based WPT, this consists in estimating the mutual inductance of the coils. If this is altered, it is assumed that an intermediate object is in the area of conflict [10]. As explained in [18], this reasoning must be carefully applied as it is difficult to determine whether the changes to this electrical parameter are due to intermediate objects or coil misalignment. The Living Object Protection (LOP) algorithms are based on FOD in order to make WPT safe.

References

1. Energous WattUp® Wire-Free Charging Technology. http://energous.com/
2. Energy Research | Navigant Research. https://www.navigantresearch.com/
3. IHS Markit | Leading Source of Critical Information. https://ihsmarkit.com/index.html
4. Ossia: Proven Wireless Power Technology You Can Use Today. http://www.ossia.com/
5. Technology - Long Range Wireless Power Transmission | Wi-Charge.com. https://www.wi-charge.com/technology/
6. Wireless Power Products - Powercastco.com. https://www.powercastco.com/
7. Chow, J.P.W., Chung, H.S.H., Cheng, C.S.: Online regulation of receiver-side power and estimation of mutual inductance in wireless inductive link based on transmitter-side electrical information. In: 2016 IEEE Applied Power Electronics Conference and Exposition (APEC), pp. 1795–1801. IEEE (2016). https://doi.org/10.1109/APEC.2016.7468111.http://ieeexplore.ieee.org/document/7468111/
8. Dai, J., Ludois, D.C.: A survey of wireless power transfer and a critical comparison of inductive and capacitive coupling for small gap applications. IEEE Trans. Power Electron. **30**(11), 6017–6029 (2015). https://doi.org/10.1109/TPEL.2015.2415253, http://ieeexplore.ieee.org/document/7064773/
9. Gibbs, Y.: NASA Dryden Fact Sheets - Beamed Laser Power (2015). https://www.nasa.gov/centers/armstrong/news/FactSheets/FS-087-DFRC.html

10. Jeong, S.Y., Thai, V.X., Park, J.H., Rim, C.T.: Self-inductance-based metal object detection with mistuned resonant circuits and nullifying induced voltage for wireless EV chargers. IEEE Trans. Power Electron. **34**(1), 748–758 (2019). https://doi.org/10.1109/TPEL.2018.2813437, https://ieeexplore.ieee.org/document/8309279/

11. Jin, K., Zhou, W.: Wireless laser power transmission: a review of recent progress. IEEE Trans. Power Electron. **34**(4), 3842–3859 (2019). https://doi.org/10.1109/TPEL.2018.2853156. https://ieeexplore.ieee.org/document/8404085/

12. Kalwar, K.A., Aamir, M., Mekhilef, S.: Inductively coupled power transfer (ICPT) for electric vehicle charging A review. Renew. Sustain. Energy Rev. **47**, 462–475 (2015). https://doi.org/10.1016/J.RSER.2015.03.040, https://www.sciencedirect.com/science/article/pii/S1364032115001938

13. Kim, H.J., Hirayama, H., Kim, S., Han, K.J., Zhang, R., Choi, J.W.: Review of near-field wireless power and communication for biomedical applications. IEEE Access **5**, 21,264–21,285 (2017). https://doi.org/10.1109/ACCESS.2017.2757267, http://ieeexplore.ieee.org/document/8052089/

14. Kisseleff, S., Chen, X., Akyildiz, I.F., Gerstacker, W.H.: Efficient charging of access limited wireless underground sensor networks. IEEE Trans. Commun. **64**(5), 2130–2142 (2016). https://doi.org/10.1109/TCOMM.2016.2550435, http://ieeexplore.ieee.org/document/7447753/

15. Lu, K., Nguang, S.K., Ji, S., Wei, L.: Design of auto frequency tuning capacitive power transfer system based on class-E2 dc/dc converter. IET Power Electron. **10**(12), 1588–1595 (2017). https://doi.org/10.1049/iet-pel.2016.0655, http://digital-library.theiet.org/content/journals/10.1049/iet-pel.2016.0655

16. Massa, A., Oliveri, G., Viani, F., Rocca, P.: Array designs for long-distance wireless power transmission: state-of-the-art and innovative solutions. Proc. IEEE **101**(6), 1464–1481 (2013). https://doi.org/10.1109/JPROC.2013.2245491, http://ieeexplore.ieee.org/document/6472725/

17. Mirbozorgi, S.A., Bahrami, H., Sawan, M., Gosselin, B.: A smart multicoil inductively coupled array for wireless power transmission. IEEE Trans. Ind. Electron. **61**(11), 6061–6070 (2014). https://doi.org/10.1109/TIE.2014.2308138, http://ieeexplore.ieee.org/document/6748029/

18. Pavo, J., Badics, Z., Bilicz, S., Gyimothy, S.: Efficient perturbation method for computing two-port parameter changes due to foreign objects for WPT systems. IEEE Trans. Magn. **54**(3), 1–4 (2018). https://doi.org/10.1109/TMAG.2017.2771511, http://ieeexplore.ieee.org/document/8122030/

19. Popovic, Z.: Cut the cord: low-power far-field wireless powering. IEEE Microw. Mag. **14**(2), 55–62 (2013). https://doi.org/10.1109/MMM.2012.2234638, http://ieeexplore.ieee.org/document/6475366/

20. Sasaki, S., Tanaka, K.: Wireless power transmission technologies for solar power satellite. In: 2011 IEEE MTT-S International Microwave Workshop Series on Innovative Wireless Power Transmission: Technologies, Systems, and Applications, pp. 3–6. IEEE (2011). https://doi.org/10.1109/IMWS.2011.5877137, http://ieeexplore.ieee.org/document/5877137/

21. Thrimawithana, D.J., Madawala, U.K.: A primary side controller for inductive power transfer systems. In: 2010 IEEE International Conference on Industrial Technology, pp. 661–666. IEEE (2010). https://doi.org/10.1109/ICIT.2010.5472724, http://ieeexplore.ieee.org/document/5472724/

22. Triviño-Cabrera, A., Aguado-Sánchez, J.: A review on the fundamentals and practical implementation details of strongly coupled magnetic resonant technology for wireless power transfer. Energies **11**(10), 2844 (2018). https://doi.org/10.3390/en11102844, http://www.mdpi.com/1996-1073/11/10/2844

23. Triviño-Cabrera, A., Lin, Z., Aguado, J.: Impact of coil misalignment in data transmission over the inductive link of an EV wireless charger. Energies **11**(3), 538 (2018). https://doi.org/10.3390/en11030538, http://www.mdpi.com/1996-1073/11/3/538

24. Trivino-Cabrera, A., Ochoa, M., Fernandez, D., Aguado, J.A.: Independent primary-side controller applied to wireless chargers for electric vehicles. In: 2014 IEEE International Electric Vehicle Conference (IEVC), pp. 1–5. IEEE (2014). https://doi.org/10.1109/IEVC.2014.7056193, http://ieeexplore.ieee.org/document/7056193/

25. Yan, Z., Zhang, Y., Kan, T., Lu, F., Zhang, K., Song, B., Mi, C.C.: Frequency optimization of a loosely coupled underwater wireless power transfer system considering eddy current loss. IEEE Trans. Ind. Electron. **66**(5), 3468–3476 (2019). https://doi.org/10.1109/TIE.2018.2851947, https://ieeexplore.ieee.org/document/8408696/
26. Zhang, Z., Chau, K.T., Qiu, C., Liu, C.: Energy encryption for wireless power transfer. IEEE Trans. Power Electron. **30**(9), 5237–5246 (2015). https://doi.org/10.1109/TPEL.2014.2363686, http://ieeexplore.ieee.org/document/6928497/
27. Zhou, W., Jin, K.: Efficiency evaluation of laser diode in different driving modes for wireless power transmission. IEEE Trans. Power Electron. **30**(11), 6237–6244 (2015). https://doi.org/10.1109/TPEL.2015.2411279, http://ieeexplore.ieee.org/document/7056428/

Chapter 2
Wireless Chargers for Electric Vehicles

2.1 Types of Electric Vehicles

Due to the current environmental crisis, there is great interest in developing new trends in the sustainable transportation sector. In this context, EVs are expected to significantly decrease greenhouse gas emissions and, in turn, lead to a healthier living environment. A number of facts support this prediction. According to the United States Environmental Protection Agency, nearly 28.9% of the greenhouse gas emissions of the United States in 2017 were derived from the transportation sector [32]. Specifically, 60% of those emissions were caused by light duty vehicles and 23% was generated by medium and heavy duty trucks. A similar figure was observed in Europe for the same year. The transportation sector of the 28 European Union member states produced 27% of greenhouse gases. This contribution was mainly caused by road transportation, which generated 72% of these emissions.

Batteries are the main energy source of EVs but some types of EVs also rely on other energetic components. Thus, there may be some EVs which work with electric propulsion alone while others also employ an Internal Combustion Engine (ICE). On that basis, the Technical Committee 69 of the International Electrotechnical Commission (IEC) has established the following classification of electric vehicles running with batteries:

- Battery Electric Vehicle (BEV). Power is inserted into the drive train exclusively by means of batteries. There is no ICE system.
- Hybrid electric vehicle (HEV). This refers to vehicles with two or more types of energy source or storage, providing one of them with electrical energy. As a result of this definition, there are multiple potential configurations for HEVs. The most common is the combination of ICE and an electrical power train activated with a battery. The activation of each system depends on the driving status. So, in low-speed conditions, the HEV relies on the electrical component whereas it opts for the ICE during high-speed driving. Under some circumstances, both systems can work together to increase the performance.

© Springer Nature Switzerland AG 2020
A. Triviño-Cabrera et al., *Wireless Power Transfer for Electric Vehicles:*
Foundations and Design Approach, Power Systems,
https://doi.org/10.1007/978-3-030-26706-3_2

Table 2.1 Battery characteristics of different EVs

Model	Vehicle type	Battery size (kWh)	Energy available (kWh)	Range (km)	Energy consumption (kWh/km)
Nissan Leaf	EV	39.5	37	240	0.165
Toyota RAV4	EV	41.8	32.18	160	0.20
Volkswagen Golf-e	EV	35.8	32	190	0.168
BMW i3	EV	33	27.2	200	0.13
Tesla Model 3 standard range	EV	55	46	310	0.148
Tesla Model S performance	EV	100	95	510	0.186
Audi e-tron	EV	95	83.6	360	0.232
Chevrolet Volt	PHEV	17.1	13.7	64	0.21
Toyota Prius	PHEV	8.8	7	40	0.175
Mitsubishi Outlander	PHEV	13.8	11	37	0.29
BMW 530e	PHEV	9.2	8	34	0.237

- Plug-in Hybrid Electric Vehicle (PHEV). This type of vehicle mainly uses the electrical power train to run. When needed (for instance when the battery level is too low), it switches to the ICE. Thus, the role of the ICE is to extend the range. The battery can be charged directly when the EV is connected to the electrical grid.

The batteries of the vehicle vary according to the manufacturer. The Table 2.1 shows the main properties of the batteries of the most popular EVs.

2.2   Battery Technology in EVs

As can be observed, BEVs, HEVs and PHEVs rely partially or completely on a battery in order to operate. However, their capacities differ in such a way that batteries in BEVs have a greater capacity than those used in PHEVs. In addition, batteries in PHEVs also have a greater capacity than those used HEVs. For example, Tesla Model 3 Standard Range is a BEV with a 55-kWh battery.

Batteries are the main storage element used in EVs. These elements make it possible to store energy in chemical form and convert it to electrical energy when required. The characteristics of the batteries, such as the density of energy and power, are defined by the technology used. Although many of them have characteristics that meet the criteria of electric vehicles, power is usually a limiting factor for some tasks such as acceleration and regenerative braking. This limitation can be mitigated by the inclusion of other technologies such as supercapacitors.

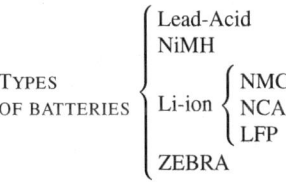

Fig. 2.1 Types of batteries used in EVs

The battery technology used in electric vehicles has evolved over time, especially in recent years with the emergence of large vehicle manufacturers in this market. The diagram in Fig. 2.1 illustrates the classification of the batteries used in EVs.

Lead-acid batteries were the first type of battery used to start the internal combustion engines in vehicles. Some manufacturers, such as Toyota and General Motors, have tested this technology in BEVs. However, the low energy density of this kind of battery does not make them suitable for pure EVs.

ZEBRA batteries, also known as molten salt batteries, have been used for some vehicle concepts and urban bus models [26]. These batteries have a good energy density, but they need to operate at high temperatures (between 270 and 350 °C). This restriction means that this technology is only viable in vehicles with a continuous operation in order to maintain the working temperature.

NiMH technology has been widely used in the market [12]. Despite the low efficiency of this technology and a slightly higher weight than others, its good energy and power density combined with its simplicity, low cost and useful lifetime make it a good solution for HEVs and PHEVs.

However, Lithium-ion or simply Li-ion batteries are the market leaders thanks to their electrical features. Within this group of batteries, we find a wide variety of different types. The types of Li-ion batteries vary according to the specific chemical combination found at the anode and the cathode. Although the combination most widely used in consumer applications is Lithium-Cobalt Oxide (LCO), its use does not extend to electric vehicles due to safety concerns. Instead, the most common solutions for automotive applications are lithium-nickel-manganese-cobalt (NMC), lithium-nickel-cobalt-aluminum (NCA) and lithium-iron phosphate (LFP). Figure 2.2 shows a comparison of the main electrical properties of Li-ion batteries in terms of lifespan, specific energy, power, safety and performance [29]. It is important to note that Lithium-ion batteries suffer from degradation, that is, their capacity diminishes in each charging/discharging cycle.

NMC batteries are used in EV applications as they provide a high energetic density. This means that their weight can also be reduced. In addition, they offer a reasonable performance in terms of efficiency, safety, lifespan and cost. BMW i3, Nissan Leaf and Chevrolet Bolt use this type of battery.

NCA batteries have traditionally been more expensive than the NMC batteries. However, Tesla and Panasonic have been working together to reduce the amount of

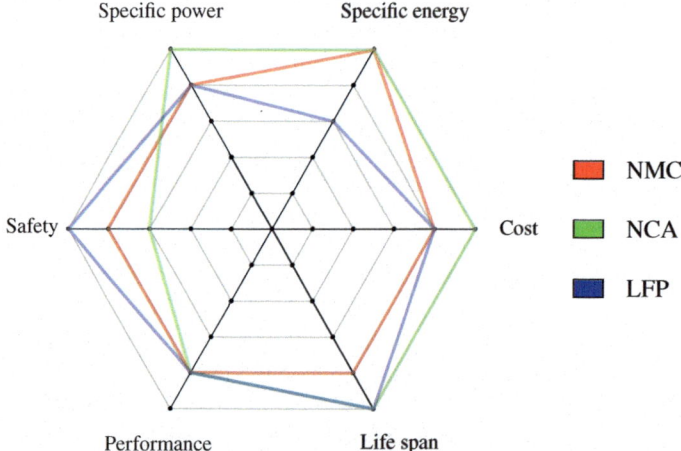

Fig. 2.2 Characteristics of Li-ion battery technologies

Table 2.2 Comparative chart of electric vehicle batteries

Type of battery	Nominal voltage (V)	Energy density (Wh/kg)	Specific power (W/kg)	Life cycle	Self discharge (% per month)	Operating temperature (°)	Production cost ($/kWh)
Lead acid (Pb-acid)	2.0	35	180	1000	<5	−15 to +50	60
Nickel-metal hydride (NiMH)	1.2	70–95	200–300	<3000	20	−20 to +60	200–250
ZEBRA	2.6	90–120	155	>1200	<5	−245 to +350	230–345
Lithium-ion (Li-ion)	3.6	118–250	200–430	2000	<5	−20 to 60	150
Lithium-iron phosphate (LFP)	3.2	120	2000–4500	>1200	<5	−45 to 70	350

cobalt required for this type of electrical storage. As a result, their prices have dropped significantly, making these batteries the cheapest on the market at present [31].

In contrast, LFP batteries are mostly used in heavy vehicles (e.g. buses), where the weight of the batteries is almost negligible. In these scenarios, the low energetic density of LFP batteries is not relevant but they contribute with a long lifespan and a safe operation.

Table 2.2 summarises the main electrical features of the batteries discussed above [16].

When developing a charging system, establishing a perfect battery model can be complex and unnecessary. Nevertheless, a vehicle battery can be modelled in different ways. Specifically, in a charge mode, the most popular model is as an equivalent resistance. The value of the resistance depends on the specifications of the battery as regards the power and voltage/current specifications.

2.2.1 Battery Management System

The most common batteries in EVs are Lithium-ion ones. This type of battery presents a reasonable energetic density and a low self-discharge rate while supporting a higher voltage in each cell. Despite these advantages, the charge and discharge process of a Lithium-ion cell must be controlled. Over-current, over-voltage or over-charging may negatively impact on the battery lifetime and even cause some safety issues such as fires or explosions [19]. Taking this factor into account, a Battery Management System (BMS) is incorporated into the vehicle to ensure that the charge/discharge process is accomplished in compliance with safety requirements. For this function, the BMS monitors several parameters related to the battery such as voltage, temperature, State of Charge (SoC), State of Health (SoH), input current and output current. With some measurements and estimations, the BMS can activate the protection circuits as needed to avoid peaks of currents, over-voltages, over-heating or abrupt discharge which will drastically reduce the battery life span. Cutoff Field Effect Transistors (FETs) allow for the controllable connection and disconnection of the battery. When any condition occurs with the potential to damage the battery, cutoff FETs completely disconnect the battery from other electrical circuits as a protection measure.

The BMS controls the current and the voltage in each of the battery components. These components are referred to as cells. Cells store energy in an electrochemical format that can be easily transformed to electrical energy by means of a reduction-oxidation process. Figure 2.3 shows a diagram of a typical BMS in which the same controller is responsible for the charge and discharge of this storage element. Some implementations opt for a separate controller for each process.

In addition to protection, the BMS is also responsible for cell monitoring and cell balancing. These functions are necessary because battery lifetime is prolonged when the cell voltages are similar.

Monitoring systems measure the voltage in each cell. With these measurements, the BMS can act in order to achieve cell balancing. Cell balancing can be accomplished through two methods. The first method, known as passive or resistor based

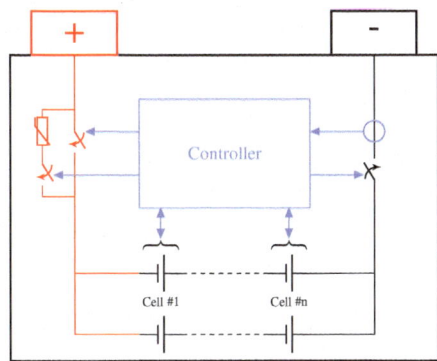

Fig. 2.3 Diagram of a BMS

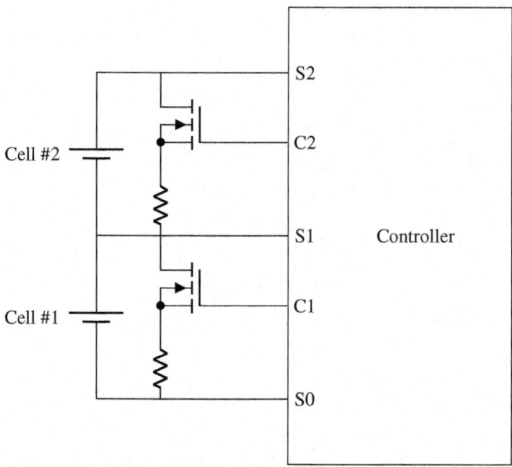

Fig. 2.4 Passive cell balancing set-up using by-pass FETs

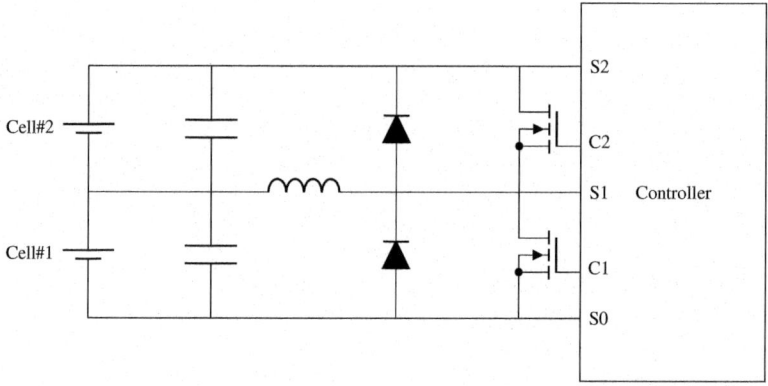

Fig. 2.5 Active cell balancing set-up using by-pass FETs

balancing, consists in discharging the cells that have an excess of voltage with a
dissipative element. By-pass FETs may be inserted in the BMS to achieve this goal,
as shown in Fig. 2.4. Because of its simplicity, it is a common implementation in
low-cost solutions. The main disadvantage associated with the passive strategy is
the energy loss incurred in the cell balancing process and the time required for this
operation. A more complex solution for cell balancing is achieved through the active
approach. Following this method, the energy is distributed among the cells in order
to balance the energy that they store. This operation can be performed during the
charge and discharge processes. EVs are usually designed with active cell balancing
(Fig. 2.5).

Finally, the BMS is also responsible for exchanging information about the battery
status with other electrical components to which it is connected. The communication interface could be MODBUS, I2C, SPI or IEEE 802.16. With this information,

the charge/discharge process is adjusted according to the real-time data in order to comply with the electrical limits of the battery and, in turn, prolong the battery lifetime.

2.3 Charging Modes in EVs

The BMS controls the energy distribution among the cells of the battery focusing on the internal aspects of this component. With a complementary view, the charging modes deal with the battery as a unity which can receive a current and/or voltage that is distributed internally by the BMS. The batteries can be charged in different ways. In this context, Society of Automotive Engineers (SAE) defines the following three charging levels in its Standard J1772 [4]:

- Level 1. It provides charging through an AC plug at 120 V, which corresponds to a standard household plug. Thus, it does not require prior installation of any specific electrical equipment to charge the vehicle.
- Level 2. Charging is achieved through an AC plug at 240 V and 40 A.
- Level 3. In contrast to previous levels, here the charging process is performed using DC current instead of AC current. The output voltage is 480 V.

An alternative classification of the methods to charge an EV battery is provided by the IEC in its standard 61851 [3]. Specifically, the Commission identifies four modes:

- Mode 1. This provides AC charging with a maximum voltage of 250 V and a maximum current of 16 A. This restricts the maximum power to 3.7 kW. A conventional household plug can support this connection. Due to overheating, it should not be used for long periods.
- Mode 2. Like mode 1, this is supported by a conventional plug but the cable incorporates a protection system. The protection is intended to check the power of the charge/discharge and to incorporate a communication scheme. The maximum current allowed in this mode is 32 A. The connection of the plug to the mains is Schuko whereas in the vehicle it is typically Mennekes.
- Mode 3. In contrast to the previous modes, this mode requires a specific installation connected to the grid. This system, which is referred to as wallbox, implements the control and protection functionalities. In addition, this mode incorporates an active communication channel between the charging point and the car. Considering the implementation of the control systems, it is the most appropriate equipment to install for smart grids.
- Mode 4. This provides a charge to the vehicle in a DC power transfer. A wallbox connected to the grid transforms the AC power to DC. This system also incorporates the protection and control mechanisms. It is especially intended for fast charging, with a maximum current of 400 A and a maximum power of 240 kW. Due to the particular features of the wallbox, it is an expensive item of equipment.

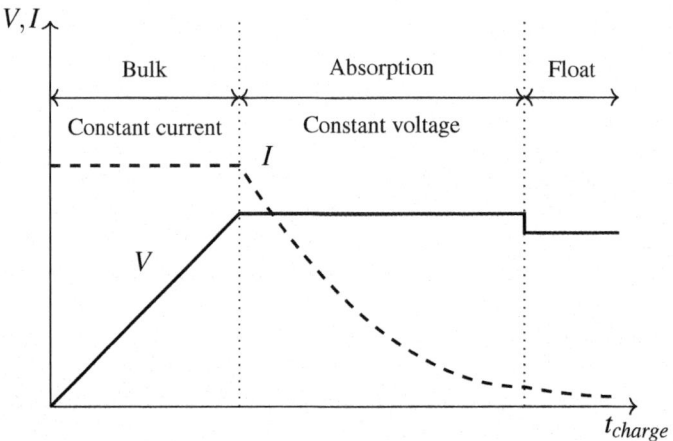

Fig. 2.6 Battery charging process in the CC–CV mode

The levels and modes described previously define how power is transferred from the grid to the vehicle. Additionally, the batteries may impose some requirements as regards the voltage and/or current during the charge process. In view of this, the charge process is accomplished in one of the following four configurations:

- Constant-Current (CC) charging. During the whole charge, a constant current rate is applied to the battery. This constant current is usually low in order to prolong the battery lifetime. The drawback of this is that it takes a long time to finish the charge. In fact, the charge ends when the charging time reaches a predefined duration, which depends on the current and the battery capacity. NiMH and Li-ion batteries can follow this approach.
- Constant-Voltage (CV) charging imposes a predefined constant voltage on charge batteries. The resulting advantage is that over-voltages and their related damages are avoided. This exclusion is expected to prolong the battery lifetime. In this charging mode, the current is decreased until it reaches a predefined value. Under these circumstances, the charge is considered to be finished.
- Constant-Current–Constant-Voltage (CC–CV) charging. The charge is developed in two phases. In the first phase, a constant current is applied. This parameter is set according to the manufacturer's recommendation and corresponds to the maximum current that the battery can tolerate without being damaged. The voltage increases during this period. When the voltage reaches its maximum, the charging mode switches to a constant voltage mode. The current in this second phase is decreased gradually. The charging is assumed to be finished when the current is lowered to a predefined value. Figure 2.6 illustrates this charging mode, which is the most common one used in EV batteries.
- Multi-stage constant-current (MCC) charging. Under this approach, the battery is charged in multiple phases or stages. In each phase, a constant current is applied. The value depends on the phase. The phase switches to a different phase, that is,

Table 2.3 Charging mode approaches [19]

Mode	Advantages	Disadvantages	Key elements
CC	Easy to implement	Capacity utilization is low	•Charging constant current rate; •Terminal condition
CV	• Easy to implement; • Table terminal voltage	Easy to cause the lattice collapse of battery	• Charging constant voltage; • Terminal condition
CC–CV	• Capacity utilization is high; • Stable terminal voltage	Difficult to balance objectives such as charging speed, energy loss, temperature variation	• Constant current rate in CC phase; • Constant voltage in CV phase; • Terminal condition
MCC	• Easy to implement; • Easy to achieve fast charging	Difficult to balance objectives such as charging speed, capacity utilization and battery lifetime	• The number of CC stages; • Constant current rates for each stage

to a different current, when the voltage reaches a threshold. This parameter is also dependent on the phase. As described in [18], the chief advantage of this mode is the short charging time it can offer. This is why it is mainly used in fast chargers. However, it is difficult to tune the parameters of each phase (current and threshold voltage) in an optimised way.

The comparison of the charging approaches is presented in the Table 2.3.

2.4 Benefits of WPT

Despite the electrical and environmental benefits of EVs, drivers are still reluctant to use them as they believe this mode of transportation could reduce their autonomy. Thus, new convenient and user-friendly methods are needed to promote the increased use of EVs. In this context, WPT is a promising technology for this form of transportation. The main advantages that WPT can offer EVs are:

• Autonomous operation. The charge/discharge can be accomplished without the driver's intervention. This autonomy is of particular interest for Vehicle-to-Grid (V2G) tasks, where the driver can participate in the electrical market without the need for any manual configuration from the user.
• Safer charge. As the driver is not required to use an electrical conductor, there is no risk of being in contact with a high electrical current. In addition, this energy transfer is safe in adverse weather conditions such as snow or rain.
• Dynamic charge. WPT extends the situations in which the EV may be charged so that it can be charged while moving or stationary for a short period. If this kind

of charge becomes widespread, it will mean that the battery of the EVs can be smaller and, consequently, the vehicle will use cheaper electrical equipment.

These advantages are relevant for all types of EVs, including cars, bicycles, buses, trains, boats and drones.

The most mature implementation of wireless chargers for EVs relies on magnetic resonance technology including commercial devices for the static mode [5] and prototypes for stationary and dynamic applications. Capacitive power transfer is also possible in the automotive sector. However, the power levels and gap are lower in comparison with the potentials of magnetic resonance chargers [8].

There are also other implementations based on alternative WPT technologies. Professor Shinohara from Kyoto University developed an experiment in which an EV was charged by means of microwaves. His research group also analysed the feasibility of a 100-kW wireless charger for EVs based on microwave energy [28]. In the far-field technologies, optical power transfer is still too immature to power EVs. As explained in [14], the power used in experiments (mainly for unmanned aerial vehicles) is restricted to 100 W with poor efficiency.

As explained previously, magnetic resonance WPT is the predominant technology in EV wireless chargers. The main challenge to address is obtaining an acceptable power transfer efficiency with a relevant gap and even when misalignment occurs. This must be done in compliance with safety considerations.

The coil design has a significant influence on system efficiency. Coils with high quality factors can improve system efficiency by up to 20% [34]. Coil geometry is another factor to consider. The influence of spiral, flat, square, and circular coils on system efficiency has been studied in [15]. For circular coils, the ratio of the primary coil and the distance to the secondary coil, r, is also a parameter for efficient designs. In [17], the authors show how system efficiency is affected by this parameter. For r lower than 0.5 the transmission efficiency will become greater than 80%. Furthermore, for r lower than 0.25, the transmission efficiency can reach 90%. High efficiencies can be also achieved by adding a ferrite core to improve the magnetic field.

Coil misalignment between the transmitter coil and the receiver coil is also another factor affecting efficiency, and this has been widely studied in [9, 25]. Lateral misalignment will inevitably occur and will reduce efficiency and transmission distance. The research challenge is how to adjust control parameters in real-time while guaranteeing the overall optimal power output.

Since there may be important inherent safety issues associated with wireless technologies for both vehicles and infrastructure, comparative further studies are needed to identify the safest architecture and operational options. Following this, acceptance testing of EVs under real-service and diverse environmental conditions is needed for operational safety assurance prior to widespread commercialisation.

2.5 WPT Operation Modes

The flexibility offered by wireless chargers increases the situations in which EVs can be charged, not only when parked but also when temporarily stopped or even in motion. As a result, the costs and the weight of the battery can be both reduced because the battery capacity can be downsized.

We can identify three wireless charging operation modes: static, stationary and dynamic. These three operation modes are illustrated in Fig. 2.7.

Static WPT occurs when the charge takes place in a specific position and the vehicle is expected to be turned off while a full charge is performed. This is the case with home chargers or those installed in car parks. The chargers can incorporate complex functions with a control feature that advises where to place the vehicle

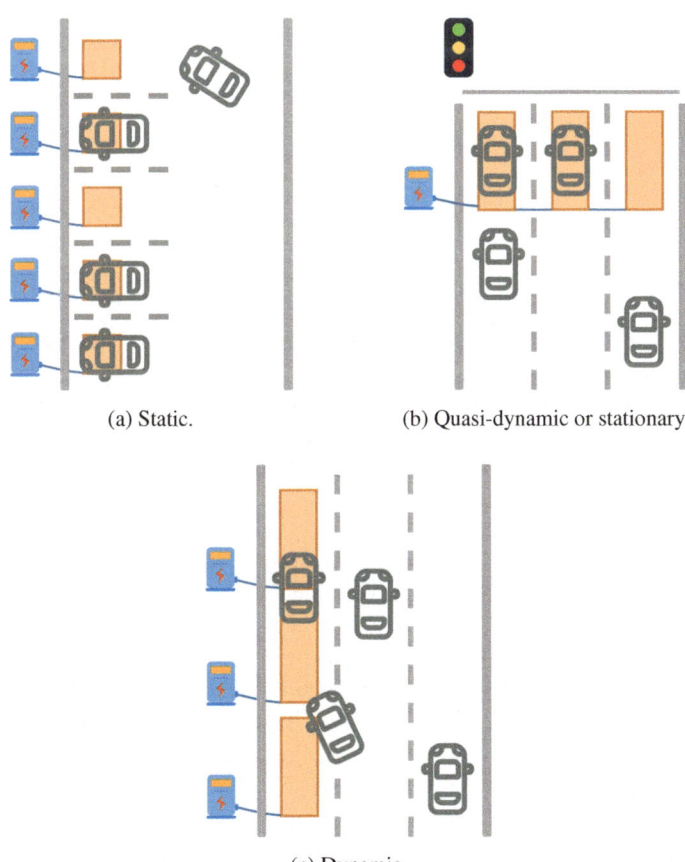

(a) Static.   (b) Quasi-dynamic or stationary.

(c) Dynamic.

Fig. 2.7 WPT operation modes

in order to prevent coil misalignment. Bombardier developed a magnetic-resonance charger for buses in Berlin providing 200 kW in a static operation mode [1].

Another type of WPT charge is the **quasi-dynamic** or **stationary**. This type of charge occurs with the following two particularities. Firstly, the vehicle is stationary but the engine is still running. Secondly, this situation holds over a short period of time that is not sufficient to reach a full charge. This type of charge is useful for public transport vehicles to receive energy when stopping at bus/tram stops or taxi ranks. This was the case with the route-63 bus trialled in Mannheim (Germany) in 2013. The bus was able to charge its battery while picking up passengers without interrupting its service to recharge the battery [2]. The bus lane was modified accordingly so as to enable this operation. Private vehicles could also be recharged during stops caused by traffic lights or conventional traffic patterns.

A relevant operation mode in WPT for EVs is the **dynamic** mode, which refers to the charge that is carried out when the vehicle is in motion, that is, when it is conventionally circulating on a paved road. Specifically, some sections of certain lanes would be equipped with electronics to enable WPT for private use or for public transport vehicles such as buses or trams. This type of charging promotes the Roadway Powered Electric Vehicle (RPEV) [22]. Dynamic charging has already been tested in some cities in South Korea (promoted by the OLEV project) and in Spain through the Victoria project. In Torino (Italy), Conductix-Wampfler has implemented a prototype to charge a bus when stopping and at the end of the bus route with magnetic resonance technology [6].

There are several initiatives already implementing wireless technologies for EV charging. Two representative examples are provided. FABRIC coordinated a study of the large-scale adoption of pure EVs in future transportation systems. This wide deployment requires mature EV technology and advanced charging solutions that provide a user experience similar to today's cars. The main challenge that FABRIC tackled was range anxiety caused by the limited range current EVs suffer from at present. In the long term, electric vehicles might be able to collect energy from the road, in a conductive or contactless mode (dynamic charge). Compared to the current paradigms of larger installed storage capacity, fast charge or switchable batteries, advanced on-road charging solutions will improve the driving range and battery lifetime of EVs as well as their energy efficiency and price, given the need for a smaller battery.

Another example is the UNPLUGGED project. This initiative investigated how the use of inductive charging of EVs in urban environments improves the convenience and sustainability of car-based mobility. In particular, it studied how smart inductive charging infrastructure can facilitate full EV integration into urban road systems while improving customer acceptance and perceived practicality. UNPLUGGED made an in-depth assessment of technical feasibility, practical issues, interoperability, user perception and socio-economic impacts of inductive charging.

2.6 Vehicle to Grid Technology

The benefits of using EVs are not only foreseen in terms of sustainability; the technology will also play an important role in smart grids. In this context, EVs can be seen as mobile generators with a practically null inertia. When required, EVs could also provide energy to the electrical network with a short response time. This function eases the management of electrical grids, particularly in those with a high penetration of renewable energy sources [33]. When an excess of renewable power generation is detected in the grid, it could be directed to charge the vehicle batteries. In an analogous way, the battery could offer ancillary services for voltage or frequency regulation [13].

Thus, the battery of an EV could have a double functionality: being charged or acting as a generator. These are known as V2G operations. The suitability of V2G operations depends on several factors related to the grid state and the battery status. Advanced algorithms take these parameters into account and provide recommendations regarding when and where to perform the charge/discharge processes [30].

According to the research agency Navigates, the V2G capacity was at 2.6 GW in 2017. This estimate represents only 1% of the global grid services market. However, this figure is growing and will reach a power capacity of 20.5 GW (4% of the grid services) in 2026.

WPT technology offers the possibility of implementing an autonomous EV charge/discharge. This capability is especially convenient for V2G operations in order to trigger the charge/discharge process in periods when the user is unable to actively participate. If these operations are to be developed further, some wireless chargers must be designed and implemented to enable both power flows: from the grid to the battery and vice-versa. Bi-directional Wireless Power Transfer (BWPT) mainly requires greater complexity in the power converters in comparison with the uni-directional wireless chargers.

2.7 Standards for EV Wireless Chargers

The promotion of EV wireless chargers requires manufacturers to guarantee their interoperability in order to ensure that any vehicle can be charged at any charging site safely and efficiently. To that end, they must comply with the requirements established by international institutions.

Two main organizations are currently working on developing specifications for EV wireless chargers. In the United States, SAE published the second version of SAE J2954 in 2017. In Europe, the IEC is the institution responsible for establishing standards through the IEC 61980. UL 2750 provides services to help product manufacturers comply with the SAE J2954 Recommended Practice, which covers safety aspects (fire, shock, and injury) related to wireless chargers. The main aspects of SAE and IEC specifications will be reviewed next.

2.7.1 SAE J2954

The Society of Automotive Engineers (SAE) is an international organisation based in the United States. Its main role consists in developing international standards for industries, especially the transport sector. Through the SAE Recommended Practice J2954 specification, this organisation aims to guarantee interoperability among uni-directional EV wireless chargers in stationary applications. For this purpose, the document defines a number of classes of EV wireless chargers and establishes a series of requirements as regards the minimum efficiency they should provide. Testing systems are also described in this Practice. They assume magnetic-resonance WPT for this type of charger.

In 2016, the first version of this document was published under the title 'Wireless Power Transfer for Light-Duty Plug-In/Electric Vehicles and Alignment Methodology'. The most recent version of the document was published in 2017 [27]. In this text, the organisation defines four types of chargers. The features of these chargers are summarised in Table 2.4.

With regard to interoperability, the requirements depend on the prior power classification. The Table 2.5 summarises the interoperability requirements.

The primary and secondary coils of WPT1 and WPT2 systems must be compatible. In contrast, WPT3 primary coils can only be compatible with WPT2 secondary coils. No requirements are specified for WPT4 coils.

In this document, the gap between the two coils is defined as the Vehicle Assembly (VA) coil ground clearance. For this parameter, SAE J2954 defines three types of operation referred to as the Z-classes. In the Z1-class, the gap ranges from 100 to 150 mm. In the Z2-class, this distance can vary from 140 to 210 mm. Finally, the Z3-class allows a distance in the range of 170–250 mm.

Table 2.4 WPT power classifications for light-duty vehicles

	WPT power class			
	WPT1	WPT2	WPT3	WPT4
Maximum input Volts Amps (kVA)	3.7	7.7	11.1	22
Minimum target efficiency at nominal x, y alignment	>85%	>85%	>85%	TBD
Minimum target efficiency at offset position	>80%	>80%	>80%	TBD

Table 2.5 Interoperability by power class

		Secondary coils			
		WPT1	WPT2	WPT3	WPT4
Primary coils	WPT1	*Required*	*Required*	Optional	Optional
	WPT2	*Required*	*Required*	Optional	Optional
	WPT3	Optional	*Required*	*Required*	Optional
	WPT4	Optional	Optional	Optional	*Required*

Table 2.6 Positioning tolerance requirements for Test Stand VAs and Product VAs

Offset direction	Value (mm)
ΔX	± 75
ΔY	± 100
ΔZ	$Z_{nom} - \Delta_{low} - > Z_{nom} - \Delta_{high}$

After presenting its classification of the EV wireless chargers, the document goes on to define some operational parameters. For these, the frequency is set to 85 kHz but can range between 81.38 and 90 kHz. The ability to cope with misalignment is also specified, as shown in Table 2.6.

In addition to variations in X, Y and Z, the primary coil and the secondary coil must hold a relevant percentage of their mutual inductance when a Roll of up to $\pm 2\%$, a Pitch of up to $\pm 2\%$ and Yaw of up to $\pm 10\%$ occur simultaneously.

In different documents, SAE is also developing requirements for the communication systems to integrate EV wireless chargers into the utility grid. These requirements are specified in SAE J2847/6, SAE J2931/6 and SAE J2836/6.

2.7.2   IEC 61980

The IEC Technical Committee 69 is working on defining a standard for wireless charge for EVs. This work is focused on the development of three parts of the IEC 61980 document. These parts are:

- Part 1: General Requirements. Published in 2015, it establishes general definitions for the concepts involved in EV WPT systems.
- Part 2: Communication between EVs and infrastructure with respect to wireless power transfer. It covers specific requirements for communication between the EV and the WPT system.
- Part 3: Specific requirements for the magnetic field power transfer systems. It addresses precise requirements for the magnetic field generated by wireless chargers in EVs in order to comply with the equipment installed in the vehicle. In this part, the 85 kHz band (81.38–90 kHz) is specified as the system frequency for passenger cars and light duty vehicles.

2.7.3   ISO 19363

The International Organization for Standardization (ISO) has developed the technical specifications 19363 under the title 'Electrically propelled road vehicles—Magnetic field wireless power transfer—safety and interoperability requirements'.

In this document, we can observe that the ISO intends to incorporate the specifications established by both of the previous institutions. As such, it also sets the 85 kHz band with a range from 81.38 to 90 kHz as the system frequency for passenger cars and light duty vehicles.

2.8 Wireless Power Transfer Market Perspectives

For transportation applications, recent market forecasts indicate an optimistic outlook for wireless EV charging markets. A combination of factors such as consumer aversion to the inconvenience of plug-in cables, the need for charging infrastructure, and features such as efficiency and reliability are expected to boost overall demand in the wireless EV charging market.

Currently, the largest market for wireless transfer applications is North America whereas the Asia Pacific region will become the dominant market for wireless EV charging systems in 2025. Demand in the European Union is expected to grow rapidly over the period 2020–2024. The market has experienced a huge increase in the supply of wireless EV chargers from start-ups, reputable technology providers and car manufacturers. The dynamic charging sector is estimated to be the fastest-growing wireless EV charging market in the period 2020–2025 [20, 23].

Wireless EV technology providers are entering into partnerships with EV manufacturers to develop business models, demonstrate commercial reliability and evaluate market niche viability. The maturity of the technology varies among providers competing to launch initiatives ranging from development and prototype test/evaluation to some in-service deployment.

Next, we will present a review of a number of technology providers and car manufacture solutions for wireless EV chargers.

2.8.1 WiTricity

WiTricity was created in 2007 as a start-up at MIT. It develops wireless power solutions using magnetic resonance technology. WiTricity is one of the leading technology providers in the wireless application market. It works with EV car manufacturers and suppliers to deploy EV wireless charging.

Following its recent acquisition of the Qualcomm Halo IP portfolio, WiTricity reinforced its role as a reference provider of wireless charging technology to car manufacturers.

WiTricitys 3.6–11 kW EV charging products show an overall system efficiency of 90–93%. WiTricity works actively with global standardisation agencies such as SAE International and IEC/ISO. The WiTricity EV wireless products offer several charging rates varying from 3.6 to 11 kW to meet different EV battery needs and with a single system design; this means that WiTricity products can charge vehicles from low to high ground clearance cars.

Witricity is a member of the AirFuel Alliance. This consortium is formed by industry leaders focused on fostering the transition to wireless power technology. This association is working on wireless charging standards including magnetic resonance to improve user experience. Its goal is to establish a holistic environment for developing inter-operability and open standards that help accelerate innovation and drive the marketplace forward.

2.8.2 Qualcomm

Qualcomm bought its "Halo" tech from the University of Auckland in 2011. It currently forms the backbone of its dynamic electric vehicle charging test track. Qualcomm Halo technology uses magnetic resonance to couple power from the electric infrastructure to the EV pad that is part of the vehicle charging system. A Qualcomm Halo system can also transfer energy from the electric vehicle battery to the electricity grid, using Vehicle to Grid technology.

This technology offers a power rate ranging from 3.3 to 20 kW at an overall efficiency greater than 90%. Qualcomm technology uses patented innovative technology that enables highly efficient power transfer even when the charging pads are not completely aligned.

In February 2019 Witricity acquired Qualcomm Inc.'s Halo wireless charging business. This agreement covers more than 1,500 patents and applications related to wireless charging. WiTricity will incorporate Qualcomm Halo into its own wireless charging operations.

2.8.3 EVATRAN

The Plugless Level 2 EV Charging System, developed by Evatran, is a wireless EV charging system successfully operating on the market. In 2011, Evatran began field trials for Plugless with installations across the USA including pilots at Google, SAP and Hertz. Since shipping its first production system in early 2014 Plugless units have provided more than 1 million trouble-free charge hours. Partnership clients have included the United States Department of Energy, through its contract with Oak Ridge National Laboratory and the Swedish Energy Agency. In 2016 VIE and Evatran established a joint venture to meet demand from OEMs manufacturing EVs for the Chinese market. In December 2016 the company demonstrated a functioning 7.2 kW-production plugless system charging a BMW i3 across 254 mm of clearance.

2.8.4   Eaton HEVO

HEVO's vision focuses on creating the global wireless charging standard for EVs that provides users with the charging experience they are demanding: the ability to simply park and power up. By offering a wireless charging option for electric vehicles, HEVO offers a safe and effective method of charging EVs that eliminates the hazards and inconveniences associated with plug-in charging.

In December 2013, Eaton Corporation announced the commercial availability of a scalable (200 kW to 1 MW) HyperCharger for fast charging hybrid and electric buses and trucks. Press articles claim it has already been installed in Tallahassee, Florida, Worcester, Massachusetts and Stockton, California, for use with Proterras EcoRide bus. It appears that 8 en-route charges have now extended the electric bus range to 240 miles per day.

2.8.5   Momentum Dynamics

Momentum Dynamics develops high-power, efficient wireless power transfer technology for the automotive and transportation industries. The technology is based on the scientific principle of magnetic resonance WPT. The Momentum Charger enables all classes of electric vehicles to be charged without supervision and under all weather conditions.

The capacity for Momentum's inductive charging technology for in-transit bus applications has been proven effective at power levels of over 200 kW, and the system is capable of delivering 450 kW. Field trials are underway, with systems that could transmit 30 kW of power across a 12-in. air gap in rain or snow.

2.8.6   Car Manufacturer Wireless Chargers

Wireless charging is now moving definitively into the mainstream automotive market. In recent months, top car manufacturers have issued press releases confirming their adoption of the technology. A brief review of some of these is provided below:

- **BMW**. BMW is bringing wireless resonant charging to several models. The wireless charger operates at 3.2 kW with an efficiency of 85%. This form of power supply to the high-voltage battery is convenient for customers and involves a charging time of around 3.5 h.
- **Audi**. Audi is set to launch its A8 L e-tron (PHEV) wireless charging platform. This has been made possible thanks to a plate connected to an industrial socket. This becomes an inductive charging station with an area of around 90 x 70 cm^2 and a height of 7 cm, weighing around 40 kg.

- **Mercedes-Benz**. Mercedes-Benz plans to launch its first wireless charging system for its electrified cars in 2019. Whereas Mercedes fastest chargers currently support up to 22 kW depending on the infrastructure [21], the wireless charging system will be limited to 3.6 kW, at least initially. The German company's first vehicle to incorporate this technology will be the S-class PHEV.
- **Hyundai-KIA**. Although the Hyundai Motor Company has not presented any product to perform wireless charging, they have registered a number of related patents. Some of these are [10], entitled "Electromagnetic field controlling system and method for vehicle wireless charging system", and [11], entitled "EV wireless charging adjustable flux angle charger". Part of this development has been coordinated by the subsidiary Hyundai-Kia America Technical Center, Inc. (HATCI) and its partner Mojo Mobility, Inc., a wireless power technology company, which have announced the joint development of a wireless charger with a charging power of 10 kW and an efficiency greater than 85% [24].
- **Continental**. Continental is not an automobile manufacturer. However, this company, known throughout the world as one of the world's largest suppliers of tyres, manufactures a large number of components for the automotive industry. Its development objectives include wireless charging technology, for which they have announced an 11-kW charger and a highly accurate vehicle approach system that successfully reduces misalignment and therefore improves transfer performance [7].

2.9 Generic Structure of the EV Wireless Charger

The most widespread technology used for wireless chargers for EVs is based on magnetic-resonance technology, and the generic structure of this type of charger is presented in Fig. 2.8.

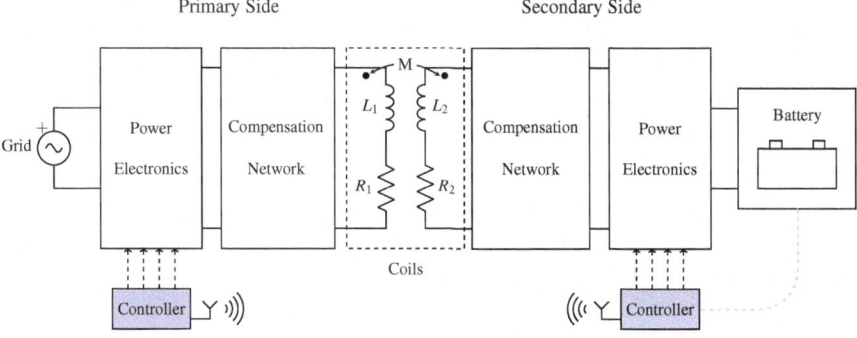

Fig. 2.8 Generic structure of an EV wireless charger

The structure of the wireless charger is divided into two main parts: the primary and the secondary side. The primary side is composed of the elements that connect the utility grid to the coil installed in the pavement. The secondary side connects to the components placed in the vehicle.

Two pairs of coils constitute the basis of this type of wireless charger. They are referred to as the primary/transmitter coil and the secondary/receiver/pick-up coil. Based on magnetic resonance technology, the primary coil is excited with a time-varying current. As a result, a magnetic field is generated. When the magnetic field traverses the secondary coil, a voltage is induced in its terminals. The geometry of the coils and its materials impact on how much of the magnetic field is used to induce voltage. These coil-related aspects are described in Chap. 3.

The system operates on resonance in order to maximize the power transfer. To achieve this, the coils are connected to two compensation networks on each side. The topology and configuration of the compensation network mainly affects its capability to cope with coil misalignment and variations in the operational frequency. Chapter 4 gives an analysis of the most relevant compensation networks.

Power converters are also necessary as the frequency provided by the utility grid is not sufficient to ensure a suitable rate of magnetic flux change. The power converter on the primary side elevates the frequency of the grid whereas the power converter on the secondary side rectifies the AC signal transmitted between the coils to a DC signal appropriate for charging the vehicle battery. The magnetic field involved in this WPT usually maintains a frequency in the range of dozens of kHz. As previously noted, the relevant international organisations are developing standards recommending that this parameter should be 85 kHz. Moreover, power electronics become useful for adjusting the power, voltage and current to the requirements of the battery. Chapter 5 makes an in-depth review of the power converters of EV wireless chargers. This chapter also includes a description of the control techniques applied to this type of device. The correct performance of the wireless charger requires the constant monitoring of some electrical parameters (e.g. the battery status informed by the BMS). Once these data are processed, the controllers must adjust the configuration of some devices. For instance, the controller may act on the switching times of the power converters' transistors.

When analysing the coils and compensation networks, it is advisable to simplify the power converters. Some equivalent circuits are derived from this reduction. On the primary side, the power converters are substituted by an AC voltage source. The amplitude and frequency of this source corresponds to the first harmonic of the converter's output signal.

The simplification of the power converters on the secondary side depends on whether the charger is uni-directional or bi-directional. For a uni-directional charger, the secondary power converter and the battery are modelled as a load resistance. The value of this resistance is related to the charge power and the voltage required by the battery. Alternatively, in a bi-directional charger the battery is conventionally modelled as DC voltage source. Figure 2.9 depicts the equivalent circuit for both types of chargers.

The simplification of each power converter is explained in more detail in Chap. 5.

(a)

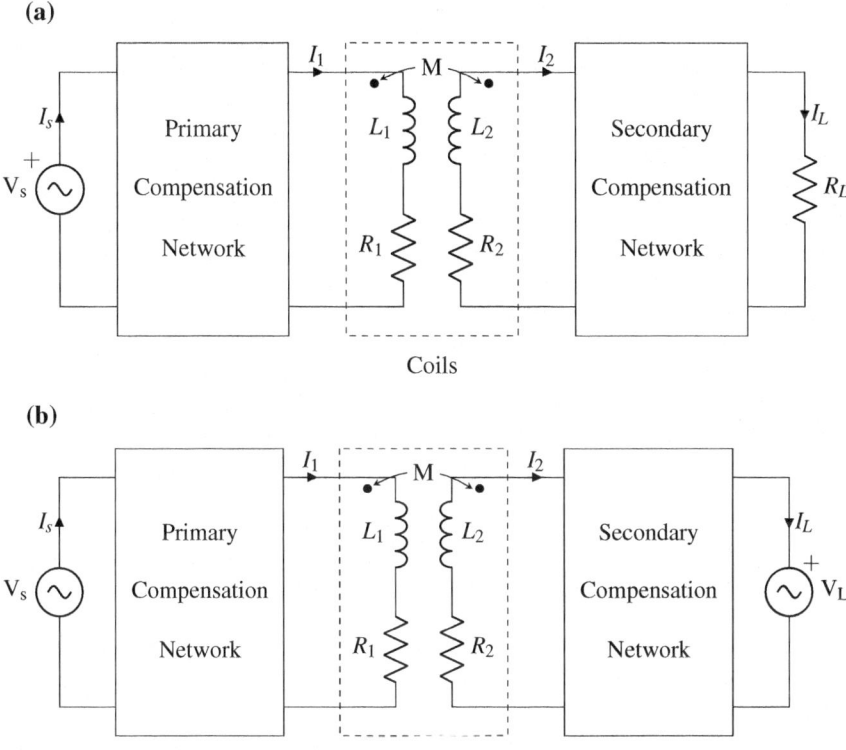

(b)

Fig. 2.9 Equivalent circuits for **a** charge mode in uni-directional and bi-directional chargers and **b** for discharge mode in bi-directional chargers

References

1. BOMBARDIER PRIMOVE to Provide Wireless Charging and Battery Technology to Berlin - Bombardier. https://www.bombardier.com/en/media/newsList/details.bombardier-transportation20150318ebusberlinabsommerfaehrtdielini.bombardiercom.html?filter-bu=tran
2. Electric Buses Test Wireless Charging in Germany | WIRED. https://www.wired.com/2013/03/wireless-charging-bus-germany/
3. IEC 61851-1:2017 | IEC Webstore. https://webstore.iec.ch/publication/33644
4. J1772: SAE Electric Vehicle and Plug in Hybrid Electric Vehicle Conductive Charge Coupler - SAE International. https://www.sae.org/standards/content/j1772_201210/
5. Meet Plugless | The Wireless EV Charging Station. https://www.pluglesspower.com/
6. Wireless charging for quiet and clean public transport in Torino (Italy) | Eltis. http://www.eltis.org/discover/case-studies/wireless-charging-quiet-and-clean-public-transport-torino-italy
7. Continental Corporation: Automated Wireless Charging from Continental: Convenient and Efficient (2017). https://www.continental-corporation.com/en/press/press-releases/2017-05-31-inductive-charging-63986
8. Dai, J., Ludois, D.C.: A survey of wireless power transfer and a critical comparison of inductive and capacitive coupling for small gap applications. IEEE Trans. Power Electron. **30**(11),

6017–6029 (2015). https://doi.org/10.1109/TPEL.2015.2415253, http://ieeexplore.ieee.org/document/7064773/

9. Fernandes, R.C., de Oliveira, A.A.: Iterative design method of weakly coupled magnetic elements for inductive power transfer. In: 2013 Brazilian Power Electronics Conference, pp. 1088–1094. IEEE (2013). https://doi.org/10.1109/COBEP.2013.6785250, http://ieeexplore.ieee.org/document/6785250/

10. Hyundai Motor Company, Kia Motors Corp, Hyundai America Technical Center Inc: Electromagnetic field controlling system and method for vehicle wireless charging system (2015). https://patents.google.com/patent/US20170120757A1/en

11. Hyundai Motor Company, Kia Motors Corp, Hyundai America Technical Center Inc: Ev wireless charging adjustable flux angle charger (2016). https://patents.google.com/patent/EP3229339B1/en?oq=EP3229339B1

12. Iclodean, C., Varga, B., Burnete, N., Cimerdean, D., Jurchiş, B.: Comparison of different battery types for electric vehicles. IOP Conf. Ser.: Mater. Sci. Eng. **252**(1), 012,058 (2017). https://doi.org/10.1088/1757-899X/252/1/012058, http://stacks.iop.org/1757-899X/252/i=1/a=012058?key=crossref.39eab45305798f03f8f65392179a6747

13. Janjic, A., Velimirovic, L., Stankovic, M., Petrusic, A.: Commercial electric vehicle fleet scheduling for secondary frequency control. Electr. Power Syst. Res. **147**, 31–41 (2017). https://doi.org/10.1016/j.epsr.2017.02.019. http://linkinghub.elsevier.com/retrieve/pii/S0378779617300767

14. Jin, K., Zhou, W.: Wireless laser power transmission: a review of recent progress. IEEE Trans. Power Electron. **34**(4), 3842–3859 (2019). https://doi.org/10.1109/TPEL.2018.2853156, https://ieeexplore.ieee.org/document/8404085/

15. Kilinc, E.G., Dehollain, C., Maloberti, F.: Design and optimization of inductive power transmission for implantable sensor system. In: 2010 XIth International Workshop on Symbolic and Numerical Methods, Modeling and Applications to Circuit Design (SM2ACD), pp. 1–5. IEEE (2010). https://doi.org/10.1109/SM2ACD.2010.5672335, http://ieeexplore.ieee.org/document/5672335/

16. Kumar, M.S., Revankar, S.T.: Development scheme and key technology of an electric vehicle: an overview. Renew. Sustain. Energy Rev. **70**, 1266–1285 (2017). https://doi.org/10.1016/J.RSER.2016.12.027, https://www.sciencedirect.com/science/article/pii/S136403211631067X

17. Kurschner, D., Rathge, C., Jumar, U.: Design methodology for high efficient inductive power transfer systems with high coil positioning flexibility. IEEE Trans. Ind. Electron. **60**(1), 372–381 (2013). https://doi.org/10.1109/TIE.2011.2181134, http://ieeexplore.ieee.org/document/6111287/

18. Lee, C.H., Chen, M.Y., Hsu, S.H., Jiang, J.A.: Implementation of an SOC-based four-stage constant current charger for Li-ion batteries. J. Energy Storage **18**, 528–537 (2018). https://doi.org/10.1016/J.EST.2018.06.010, https://www.sciencedirect.com/science/article/pii/S2352152X18300513

19. Liu, K., Li, K., Peng, Q., Zhang, C.: A brief review on key technologies in the battery management system of electric vehicles. Front. Mech. Eng. **14**(1), 47–64 (2019). https://doi.org/10.1007/s11465-018-0516-8, http://link.springer.com/10.1007/s11465-018-0516-8

20. Market Research Future: Wireless Power Transmission Market Research Report - Forecast 2022 | MRFR. Technical report (2019). https://www.marketresearchfuture.com/reports/wireless-power-transmission-market-2341

21. Mercedes Benz: EQ - Charging at home (2019). https://www.mercedes-benz.com/en/eq/about-eq/charging-and-services/charging-at-home/

22. Mi, C.C., Buja, G., Choi, S.Y., Rim, C.T.: Modern advances in wireless power transfer systems for roadway powered electric vehicles. IEEE Trans. Ind. Electron. **63**(10), 6533–6545 (2016). https://doi.org/10.1109/TIE.2016.2574993, http://ieeexplore.ieee.org/document/7491313/

23. Navigant Research: The Disruptive Potential of Wireless EV Charging. Technical report (2018). https://www.navigantresearch.com/reports/the-disruptive-potential-of-wireless-ev-charging

24. New Mobility: Hyundai-Kia America Technical Center and Mojo Mobility partner on wireless EV charging system (2018). https://www.newmobility.global/technology/hyundai-kia-america-technical-center-mojo-mobility-partner-wireless-ev-charging-system/

25. Nguyen, M.Q., Hughes, Z., Woods, P., Seo, Y.S., Rao, S., Chiao, J.C.: Field distribution models of spiral coil for misalignment analysis in wireless power transfer systems. IEEE Trans. Microw. Theory Tech. **62**(4), 920–930 (2014). https://doi.org/10.1109/TMTT.2014.2302738, http://ieeexplore.ieee.org/document/6729099/

26. O'Sullivan, T., Bingham, C., Clark, R.: Zebra battery technologies for all electric smart car. In: International Symposium on Power Electronics, Electrical Drives, Automation and Motion, 2006. SPEEDAM 2006., p. 243. IEEE. https://doi.org/10.1109/SPEEDAM.2006.1649778, http://ieeexplore.ieee.org/document/1649778/

27. SAE International: Wireless Power Transfer for Light-Duty Plug-In/Electric Vehicles and Alignment Methodology (SAE TIR J2954) (2012). https://www.sae.org/standards/content/j2954/

28. Shinohara, N., Kubo, Y., Tonomura, H.: Wireless charging for electric vehicle with microwaves. In: 2013 3rd International Electric Drives Production Conference (EDPC), pp. 1–4. IEEE (2013). https://doi.org/10.1109/EDPC.2013.6689750, http://ieeexplore.ieee.org/document/6689750/

29. The Boston Consulting Group: Batteries for Electric Cars: Challenges, Opportunities, and the Outlook to 2020. Technical report (2010). http://gerpisa.org/en/node/614

30. Triviño-Cabrera, A., Aguado, J.A., Torre, S.d.l.: Joint routing and scheduling for electric vehicles in smart grids with V2G. Energy **175**, 113–122 (2019). https://doi.org/10.1016/J.ENERGY.2019.02.184, https://www.sciencedirect.com/science/article/abs/pii/S0360544219303901

31. Tsiropoulos, I., Tarvydas, D., Lebedeva, N.: Li-ion batteries for mobility and stationary storage applications. Technical report, Joint Research Centre (JRC) (2018). https://doi.org/10.2760/87175, https://ec.europa.eu/jrc/en/publication/eur-scientific-and-technical-research-reports/li-ion-batteries-mobility-and-stationary-storage-applications

32. United States Environmental Protection Agency: Sources of Greenhouse Gas Emissions (2017). https://www.epa.gov/ghgemissions/sources-greenhouse-gas-emissions

33. Wang, K., Gu, L., He, X., Guo, S., Sun, Y., Vinel, A., Shen, J.: Distributed energy management for vehicle-to-grid networks. IEEE Netw. **31**(2), 22–28 (2017). https://doi.org/10.1109/MNET.2017.1600205NM, http://ieeexplore.ieee.org/document/7884945/

34. Li, Y., Li, X., Peng, F., Zhang, H., Guo, W., Zhu, W., Yang, T.: Wireless energy transfer system based on high Q flexible planar-Litz MEMS coils. In: The 8th Annual IEEE International Conference on Nano/Micro Engineered and Molecular Systems, pp. 837–840. IEEE (2013). https://doi.org/10.1109/NEMS.2013.6559855, http://ieeexplore.ieee.org/document/6559855/

Chapter 3
Coil Design for Magnetic Resonance Chargers

3.1 Introduction

Magnetic resonance WPT is based on a pair of coils which are coupled through an air gap. The geometry and materials of this pair of elements are crucial for determining the magnetic field of the WPT and its efficiency. Additionally, the electrical parameters of the coils will determine the configuration of the compensation networks. For these reasons, coil design is one of the most important steps in the complete definition of an EV WPT charger.

The geometry and configuration of the coils must be decided in a complex design process, taking into account economical and electrical factors. Firstly, the use of specific materials to operate at high frequencies makes these coils more expensive; as such, the length of the cables used for the turns and the size of ferromagnetic elements should be carefully considered. In addition, the design of the coils is decisive for the self-inductance, their quality factors and the mutual inductance. Figure 3.1 shows the equivalent diagram of two coupled coils. In contrast to strongly coupled magnetic resonance applications, the parasitic capacitance is usually omitted in EV chargers analysis due to its low value in comparison with the capacitive elements of the compensation networks.

All the coil parameters have a direct impact on the electrical performance of the WPT system. These effects reveal themselves not only at predefined positions of the coils but also when there are coil misalignments. Some geometries are specifically designed in EV WPT applications to cope with coil displacements. It is important to note that there is a high probability that the charging and the pick-up coils are not perfectly aligned in the three operation modes identified in Chap. 2 (static, stationary and dynamic charging). As illustrated in Fig. 3.2, there are four types of misalignments: horizontal, vertical, horizontal plus vertical and angular. Angular misalignment is not common in EV wireless chargers.

Complex coil structures incorporate magnetic and/or metallic materials to control the propagation of the magnetic field involved in the WPT. Ferromagnetic mate-

© Springer Nature Switzerland AG 2020
A. Triviño-Cabrera et al., *Wireless Power Transfer for Electric Vehicles: Foundations and Design Approach*, Power Systems,
https://doi.org/10.1007/978-3-030-26706-3_3

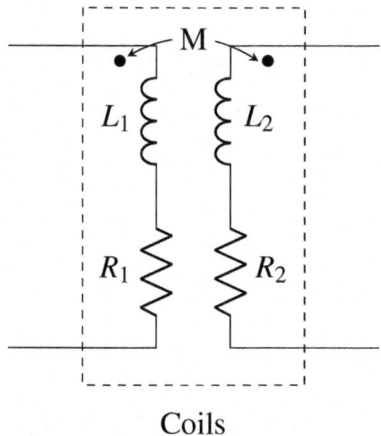

Coils

Fig. 3.1 Electric parameters of the coupled coils

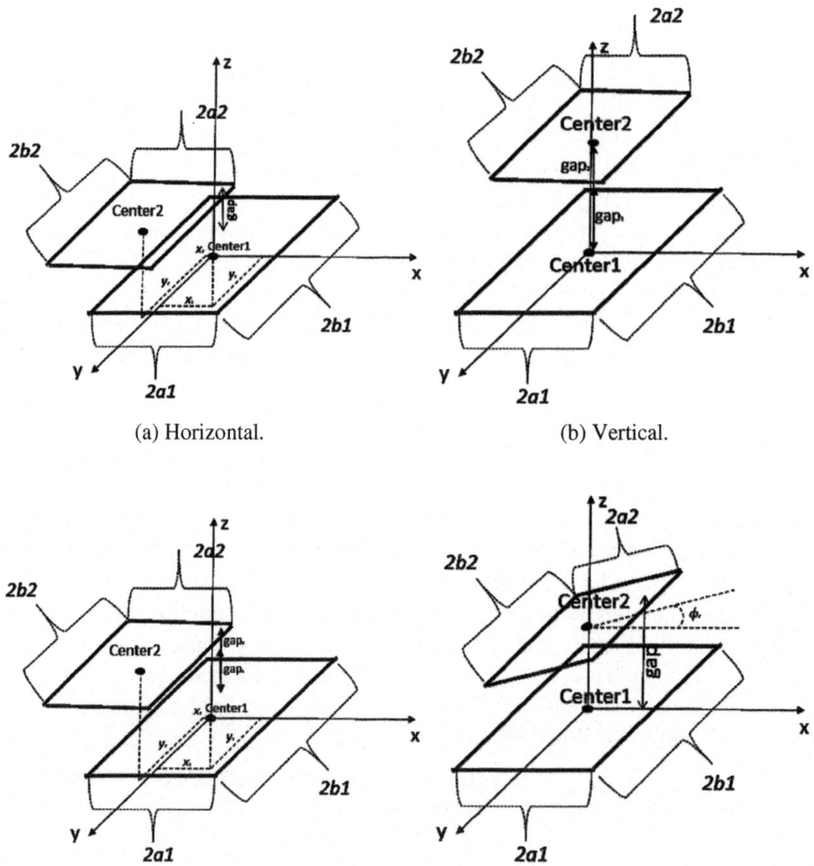

(a) Horizontal.

(b) Vertical.

(c) Verticalandhorizontal.

(d) Angular.

Fig. 3.2 Types of misalignments between a pair of coils

rials help to guide the magnetic field to the area of interest whereas the metallic elements prevent its propagation in protected zones. The inclusion of these materials is non-trivial as they alter the electrical parameters of the coils. Moreover, these modifications cannot be easily estimated by equations, and so designers usually opt for software based on Finite Element Analysis to characterise them [1–3]. The main concern with this approach is that this kind of tool requires significant computational resources. As a result, the evaluation of the coil parameters using this method takes a considerable length of time and is supported by a noticeable memory allocation.

A correct design of the coils must consider the economical and electrical features of these elements in order to determine their geometry, the materials on which they are built, their configuration (number of turns and dimensions) and the suitability of applying ferromagnetic materials or shielding.

First, this chapter describes the most common geometries used for the coils in EV wireless chargers. A review is then given of the magnetic fields generated with these coils. This study is extended through an analysis of the magnetic fields when the coils incorporate ferromagnetic materials and/or shielding.

3.2 Coil Geometry and Materials

Although there are some proposals for 3D coil designs [27], most designs in EV WPT opt for planar coils. In particular, this is the usual case for the pick-up coil. As we will see in the next subsection, the transmitter coil may be bulky in high-power applications where the ferromagnetic material is placed around the track coil.

The first implementations of EV wireless chargers were supported by simple coils. In particular, two options were most popular: circular and rectangular coils (Fig. 3.3).

For **circular coils** of N turns, their self-inductance L can be determined according to the following equation:

$$L = \mu_0 * N^2 * R * \left(\ln \frac{16 * R}{d} - \frac{7}{8} \right)$$

(3.1)

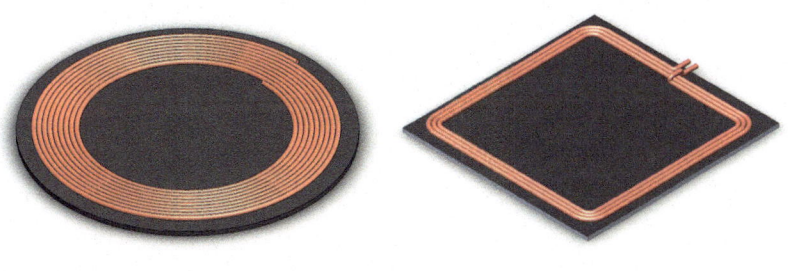

(a) Circular coil.   (b) Rectangular coil.

Fig. 3.3 Simple coil geometries for EV wireless chargers

where μ_0 is the vacuum permeability, N is the number of turns, R is the radius of the inner turn and d corresponds to the equivalent diameter of the coils. The expression for this last parameter is:

$$d = 2 * \sqrt{\frac{N * S}{\pi}} \tag{3.2}$$

where S is the cross-section of the cable. If the cable is cylinder-shaped with a radius equal to R, then $S = \pi * R^2$.

Alternatively, the self-inductance of **rectangular coils** (with dimensions $a \times b$ m^2) is computed as:

$$L = \mu_0 * \pi * N^2 * [(a+b) * \log{(4 * \frac{a*b}{d})} - a * \log{(a+\sqrt{a^2+b^2})}$$
$$-b * \log{(b+\sqrt{a^2+b^2})} + 2 * \sqrt{a^2+b^2} + (d - 2*(a+b))] \tag{3.3}$$

As compared in [14], for the same length of cable, circular geometries lead to higher self-inductance than rectangular ones.

The previous equations assume that the turns of the coils have no space between them, that is, a null pitch. However, some implementations of EV coils include a pitch, leading to the geometries shown in Fig. 3.4.

With this type of configuration, the self-inductance of a circular coil is:

$$L = \frac{\mu_0 N^2 d_{avg}}{2} \left(ln\left(\frac{2.46}{\phi} + 0.2\phi^2\right) \right) \tag{3.4}$$

where

$$D_{avg} = \frac{d_{out} + d_{in}}{2} \tag{3.5}$$

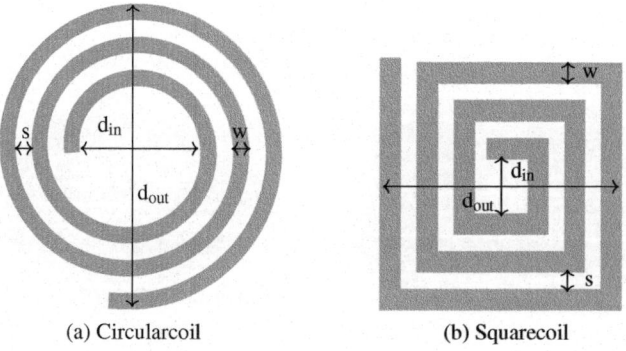

(a) Circularcoil   (b) Squarecoil

Fig. 3.4 Geometries of coils with pitch

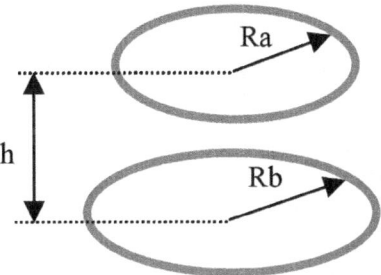

Fig. 3.5 Circular coils dimension for M calculation

$$\phi = \frac{d_{out} - d_{in}}{d_{out} + d_{in}} \tag{3.6}$$

d_{out} is the outer diameter and d_{in} corresponds to the inner diameter of the coil.

In a rectangular coil, the new self-inductance corresponds with:

$$L = 2,34 \cdot \mu_0 * \frac{N^2 \cdot d_{avg}}{1 + 2.75 \cdot \phi} \tag{3.7}$$

With regard to their mutual inductance, there are several methods for computing this parameter, such as Heumans lambda function, Bessel and Struve functions and Biot–Savart law [15].

Taking into account the configuration of the circular coils presented in Fig. 3.5, the mutual inductance M between them is:

$$M = \mu_0 N_1 N_2 \pi \sqrt{R_a R_b} \left(\frac{F_K^3}{16} + 3 \frac{F_K^3}{64} \right) \tag{3.8}$$

where

$$F_K = \sqrt{\frac{4 R_a R_b}{(R_a + R_b)^2 + h^2}} \tag{3.9}$$

When there is lateral misalignment, the mutual inductance between circular and planar coils varies.

As presented in [10, 15], the mutual inductance of rectangular coils is lower than for circular geometries. However, rectangular coils present a better tolerance to coil misalignment than circular ones. In [28], the authors show how the circular coupler incurs higher costs per kW in comparison with the rectangular geometry.

Due to their simplicity, circular and rectangular coils are still used today. However, other complex geometries have been proposed with the goal of offering a lower sensitivity to coil misalignment. They are mainly the DD, the DDQ and the bipolar coils. These three geometries are presented in Fig. 3.6.

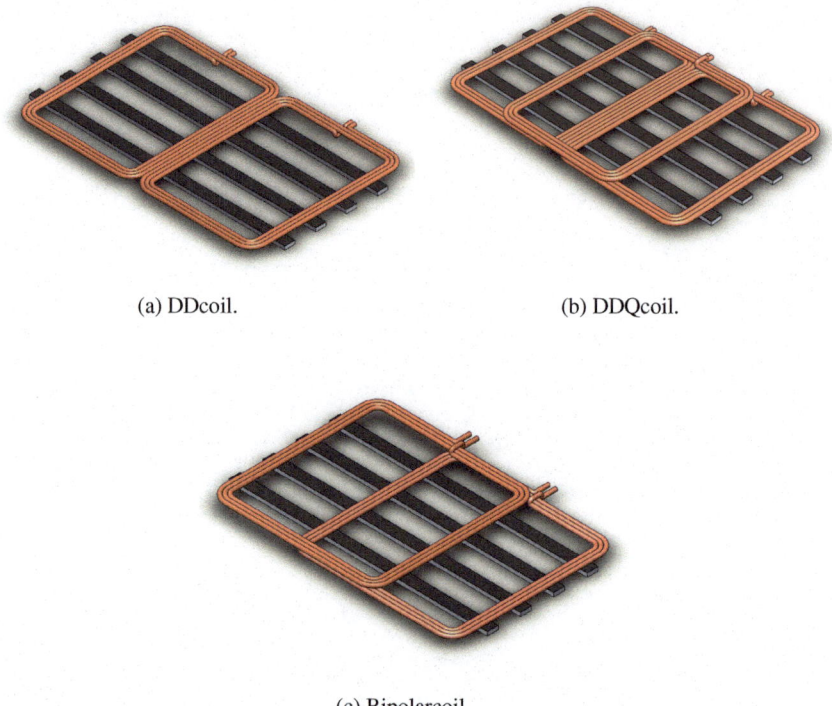

(a) DDcoil.   (b) DDQcoil.

(c) Bipolarcoil.

Fig. 3.6 DD, DDQ and bipolar coil geometries

Double-D or **DD coils** are constituted by two equal D-shaped (rectangular) sub-coils connected in parallel and with a shared side [6]. The wound in each coil is opposite so that the current in one sub-coil is clockwise whereas in the other it goes in the opposite direction. They are said to be in series from a magnetic point of view but with a parallel electrical connection. As the two sub-coils are in contact, the magnetic fields generated by one coil affect the other so that the two sub-coils are coupled.

DD quadrature or **DDQ** is built with two independent windings. The first winding follows the DD geometry. The second coil, referred to as the quadrature or Quad (Q) coil, is built overlapping half of the area of each D component. They are usually supported by a ferrite and metallic structure underneath which ensures that the DD-coil and the Q-coil are not coupled. The self-inductance of the D-coil and the quadrature coil usually differs. The design procedure can be executed independently. As explained in the following section, this topology provides great flexibility to cope with misalignments and inter-operability. However, the use of this type of coil makes the power electronics and its control more complex as there are two subsystems to control. The diagram in Fig. 3.7 shows the equivalent circuit of a secondary coil implemented with a DDQ geometry [17].

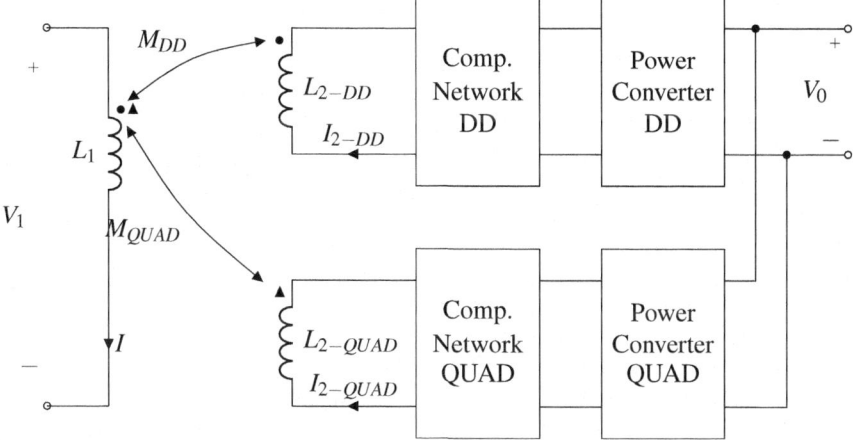

Fig. 3.7 DDQ equivalent circuit

Table 3.1 Comparison of different pad structures

Coils structure	Misalignment tolerant	Coefficient of coupling	EMF exposure	Effect of shielding on coefficient of coupling	Magnetic flux
Circular pad	Poor	Low	High	Low	Single sided
DD coil	Poor	High	Low	High	Double sided
DDQ coil	High	High	Low	High	Double sided
Bipolar pad	Medium	High	Low	High	Double sided

The basis of a **bipolar** coil again follows equivalent D-shape coils. In contrast to DD-coils, one of the coils overlaps half of the area of the other D-shape coil. This can be considered an intermediate option between DD and DDQ coils. It is able to concatenate more magnetic flux than DD coils but less than DDQ ones. However, due to its simpler geometry, the costs and losses incurred for the copper are lower compared with DDQ coils [5].

Table 3.1 extends the comparison presented in [19] regarding coil geometries.

For dynamic wireless charging applications, the primary coil (also known as the track coil) can be implemented with two different geometries as shown in Fig. 3.8. In the first approach, the track coil has a similar dimension to the pickup coil (the one installed in EVs). Several track coils are placed along a lane in order to maintain the charge process for a reasonable length of time. In contrast, the longitudinal dimension of the stretch coils are much larger than the size of the pick-up coils.

With regard to the material of the coils, it is recommended to use a wire with low resistance because reduced losses in the inductors lead to the increased overall efficiency of the wireless charger [24]. The equivalent resistance of a wire depends on its length and also on the operational frequency. Thus, we define two terms: the DC and the AC winding loss. The DC losses are mainly due to the resistivity of the conductor whereas the AC losses are caused by the eddy current effects.

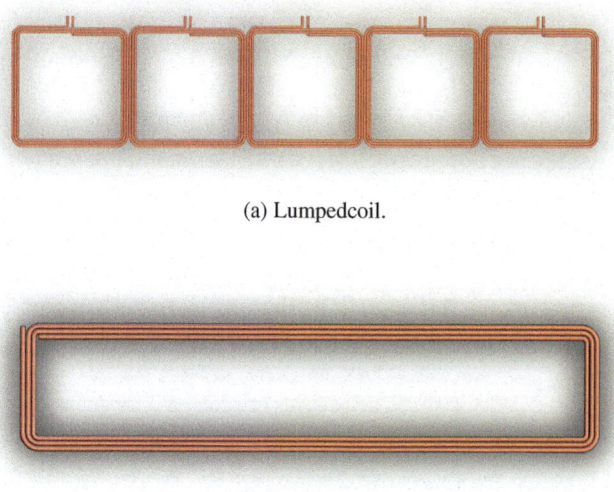

(a) Lumpedcoil.

(b) Stretchedcoil.

Fig. 3.8 Lumped and stretched coil geometries for dynamic WPT applications

The eddy current effects combine two phenomena: the skin and the proximity effects [22]. Skin effect refers to the phenomenon in which the AC current flowing in a conductor tends to concentrate near the surface. The higher the frequency, the more concentrated the current is in the 'skin'. This concentration results in a reduction of the effective cross-section of the cable. In contrast, the proximity effect is due to parallel conductors carrying current. Because of the current flowing in a conductor, a magnetic field is generated, altering the current concentration in the parallel conductors. Specifically, the current tends to concentrate in the areas furthest away from the conductor generating the initial magnetic field.

The AC winding losses reveal themselves with a non-uniform current density diminishing the effective area in which the current flows. In fact, this variation is non-linearly dependent with the frequency.

To minimize the AC winding losses, Litz wire is used in EV wireless chargers. A Litz wire conductor consists of multiple insulated parallel strands; the DC resistance of each strand is equivalent. These insulated strands are twisted to form a bundle. Then, multiple bundles are twisted again leading to a bundle of bundles. Figure 3.9 shows an image of a Litz wire section.

The goal of this type of conductor is to achieve a uniform current density. As explained in [26], we should vary the strand diameter to get uniform current density at the switching frequency and its harmonics. The bundle diameter should be adjusted to obtain this uniformity at the fundamental frequency. If these two parameters are set appropriately, the effective area of a Litz wire is much higher than for a solid-round wire. As a result, the skin effect can be ignored. Figure 3.10 illustrates how the Litz wire achieves a uniform current density for a test carried out in Ansys Maxwell.

Fig. 3.9 Litz wire

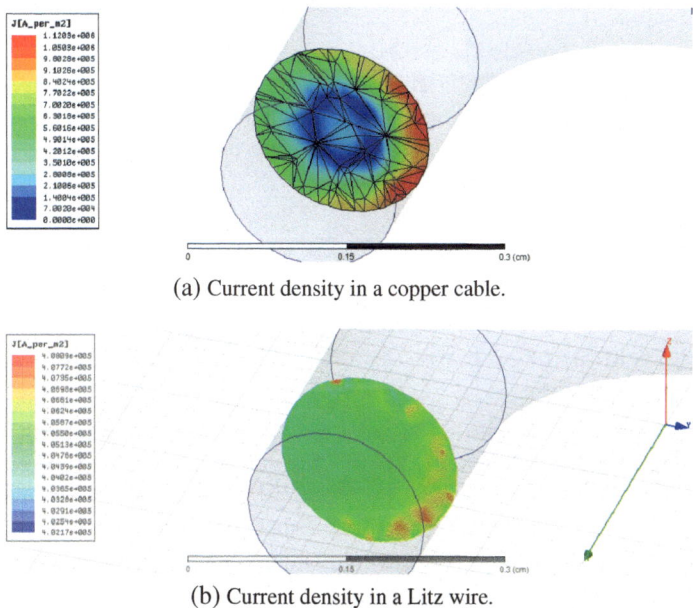

(a) Current density in a copper cable.

(b) Current density in a Litz wire.

Fig. 3.10 Illustration of the skin effect of copper cable versus Litz wire

Litz wire is classified according to the gauge of the strands, which is selected depending on frequency range and the number of strands. Table 3.2 presents the recommended strand gauge for each operation frequency range using the American Wire Gauge (AWG) as this is usually found in the literature.

Consequently, some expressions can be derived for the DC and AC resistance of a cable. In particular, for a Litz-wire conductor we find that the DC losses (R_{WDC}) are:

$$R_{WDC} = \frac{4\rho_w l_w}{k\pi d_{str}^2} = \frac{4\rho_w l_T N_l}{k\pi d_{str}^2} \tag{3.10}$$

Table 3.2 Recommended strand gauge for each operation frequency range [18]

Operation frequency range	AWG	Diameter (mm)
60 Hz to 1 kHz	28	0.3211
1 kHz to 10 kHz	30	0.2546
10 kHz to 20 kHz	33	0.1798
20 kHz to 50 kHz	36	0.1270
50 kHz to 100 kHz	38	0.1007
100 kHz to 200 kHz	40	0.0799
200 kHz to 350 kHz	42	0.0635
350 kHz to 850 kHz	44	0.0508
850 kHz to 1.4 MHz	46	0.0398
1.4 MHz to 2.8 MHz	48	0.0316

where $\rho = \frac{1}{w}$ is the conductor resistivity, and d, l_w and l_T correspond to the diameter of the wire, the total winding length and the mean turn length.

The AC resistance of a Litz wire (R_{WAC}) is approximated to [26]:

$$R_{WAC} = \frac{4\rho_w l_w}{\pi k}\left(\frac{1}{d^2} + \frac{\pi^3(5N_{ll}^2 - 1)d^4}{2880\delta^4 p^2 k^2}\right) \tag{3.11}$$

3.3 Electromagnetic Emissions of Two Coupled Coils

The determination of the electromagnetic field of a magnetic resonance system is required due to two main reasons. Firstly, it provides a mathematical tool to infer and optimise how the power transfer is accomplished. Magnetic flux leakage can be avoided with the appropriate incorporation of sophisticated materials in the charger.

Secondly, it allows engineers to comply with the safety recommendations related to the maximum magnetic emissions. This includes interference with other electrical devices (EMI) and human safety. Magnetic fields induce voltages and currents, which can alter the conventional behaviour of the body. Depending on the magnetic field intensity and the time of exposure, the effects may be temporary or permanent. Although there is no definitive conclusion on the precise correlation between magnetic field intensity and the harmful effects on the body, some research has indeed found evidence of a link. For instance, the analysis in [23] demonstrates that the increase in 1 °C of human tissue already shows some modifications in its biological functions. Children are more sensitive to magnetic radiation as the absorption capacity of their tissue is ten times higher than in adults. Long-term consequences (especially in spatial memory) were observed.

In order to prevent these harmful effects, several organisations have established the maximum levels permitted for magnetic field intensity. In the United States, the

Table 3.3 Basic restrictions for electric, magnetic and electromagnetic fields (0 Hz to 300 GHz)

Occupational exposure

Frequency range	E-field strength E (kV/m)	Magnetic field strength H (A/m)	Magnetic flux density B (T)
1–8 Hz	20	$1.63 \times 10^5/f^2$	$0.2/f^2$
8–25 Hz	20	$2 \times 10^4/f$	$2.5 \times 10^{-2}/f$
25–300 Hz	$5 \times 10^2/f$	8×10^2	1×10^{-3}
300–3 kHz	$5 \times 10^2/f$	$2.4 \times 10^5/f$	$0.3/f$
3 kHz–10 MHz	1.7×10^{-1}	80	1×10^{-4}

General public exposure

Frequency range	E-field strength E (kV/m)	Magnetic field strength H (A/m)	Magnetic flux density B (T)
1–8 Hz	5	$3.2 \times 10^4/f^2$	$4 \times 10^{-2}/f^2$
8–25 Hz	5	$4 \times 10^3/f$	$5 \times 10^{-3}/f$
25–50 Hz	5	1.6×10^2	2×10^{-4}
50–400 Hz	$2.5 \times 10^2/f$	1.6×10^2	2×10^{-4}
400–3 kHz	$2.5 \times 10^2/f$	$6.4 \times 10^4/f$	$8 \times 10^{-2}/f$
3 kHz–10 MHz	8.3×10^{-2}	21	2.7×10^{-5}

Code of Federal Regulations (CFR) established these limits. In China, the recommendations in GB9175-88 are followed. In Europe, the International Commission on Non-Ionizing Radiation Protection (ICNIRP) is responsible for determining exposure limits for electric and magnetic variable fields in the range from 1 Hz to 100 kHz. The first standard set was the ICNIRP 1998, which established the maximum levels of Specific Absorption Rate (SAR) for the body, head and legs. This document also sets limits for the induced current density in the head and trunk of the body. A new revised version of this document was published with the ICNIRP 2010, in which the limits to field exposure were modified and new maximum levels were established for the induced electric field in the central nervous system. There is also a relevant recommendation in the Institute of Electrical and Electronics Engineers (IEEE) Standard for Safety Levels with Respect to Human Exposure to RF EMF 3 kHz to 300 GHz.

The ICNIRP restrictions [12] on field exposure are summarised in Table 3.3.

Considering these restrictions, engineers working with EV wireless chargers need to estimate the magnetic field involved in these devices during the design process. As presented in [25], the intensity of the magnetic field is a time-space function that depends on the currents (their modules and phases) flowing in the two coils. In Chap. 4, we will see how these currents are affected by misalignment and the compensation networks.

As an illustrative example, we will derive the magnetic field of two rectangular coils used in a magnetic resonance charger. For this analysis, we will start with the definition of the magnetic field generated by a square coil (with dimensions $2a \times 2b$)

centered in a generic position (u, v, w) with a current I. The parameters r_1, r_2, r_3 and r_4 are defined for this configuration as follows:

$$r_1 = \sqrt{(a + x - u)^2 + (y - v + b)^2 + (z - w)^2} \qquad (3.12)$$

$$r_2 = \sqrt{(a - x + u)^2 + (y - v + b)^2 + (z - w)^2} \qquad (3.13)$$

$$r_3 = \sqrt{(a - x + u)^2 + (y - v - b)^2 + (z - w)^2} \qquad (3.14)$$

$$r_4 = \sqrt{(a + x - u)^2 + (y - v - b)^2 + (z - w)^2} \qquad (3.15)$$

The components of the magnetic field in a position (x, y, z) due to a loop of the coil are:

$$\overline{B_x} = \frac{\mu_0 I}{4\pi} \sum_{\alpha=1}^{4} \frac{(-1)^{\alpha+1}(z - w)}{r_\alpha(r_\alpha + d_\alpha)} \qquad (3.16)$$

$$\overline{B_y} = \frac{\mu_0 I}{4\pi} \sum_{\alpha=1}^{4} \frac{(-1)^{\alpha+1}(z - w)}{r_\alpha(r_\alpha + (-1)^{\alpha+1}C_\alpha)} \qquad (3.17)$$

$$\overline{B_z} = \frac{\mu_0 I}{4\pi} \sum_{\alpha=1}^{4} \left[\frac{(-1)^{\alpha+1}d_\alpha}{r_\alpha(r_\alpha + (-1)^{\alpha+1}C_\alpha)} - \frac{C_\alpha}{r_\alpha(r_\alpha + d_\alpha)} \right] \qquad (3.18)$$

where

$$C_1 = -C_4 = a + x - u \qquad (3.19)$$

$$C_2 = -C_3 = a - x - u \qquad (3.20)$$

$$d_1 = d_2 = y - v + b \qquad (3.21)$$

$$d_3 = d_4 = y - v - b \qquad (3.22)$$

Thus, the complete magnetic field due to the N turns of the coil corresponds to:

$$B(u, v, w, I) = N\sqrt{B_x^2 + B_y^2 + B_z^2} \qquad (3.23)$$

In a magnetic resonance charger, we will have two coils allowing potential misalignment as presented in Fig. 3.11. Assuming a linear magnetic medium, the magnetic field generated by the two coils (B_{MR}) is computed as the sum of the contribution of each coil individually. Considering that the currents of the primary and secondary coils are $I_1 \angle \phi_1$ and $I_2 \angle \phi_2$ respectively, the resulting magnetic field is equal to:

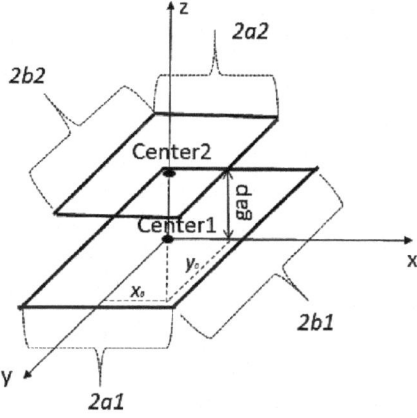

Fig. 3.11 Two coupled coils

$$B_{MR} = N_1 B(0, 0, 0, I_1) \sin (\omega t + \phi_1) + N_2 B(x_o, y_o, gap, I_2) \sin (\omega t + \phi_2)$$
$$(3.24)$$

where N_1 and N_2 are the number of turns of the primary and secondary coil respectively.

Since magnetic resonance chargers transfer power based on magnetic fields, knowing the main features of this type of field is of particular interest not only to comply with the restrictions on field exposure but also to optimise the power transfer. Specifically, the voltage induced in the secondary coil is related to the variation of the magnetic flux traversing its area. This condition implies that circular coils without angular misalignment are only able to capture the vertical component of the magnetic field. An opposite behaviour is observed in DD coils, which only use the horizontal component to induce voltage in their terminal. In contrast, DDQ coils can capture all of the available flux in the horizontal and vertical directions. These capabilities give us an insight into the ability of secondary coils to cope with certain types of misalignment. This feature is included in Table 3.1.

3.4 Use of Ferromagnetic Material

The inclusion of ferromagnetic material in a magnetic resonance charger sets a preferential path for the magnetic field so that its intensity can be greater in those areas of interest whereas the leakage magnetic flux can be diminished. The result is a more efficient power transfer. The inclusion of this kind of material in WPT systems is known as passive magnetic shielding.

When deciding which material to use in an application, two parameters must be studied: the relative permeability of the material, and its losses. Firstly, the permeability should be as high as possible to ensure low magnetic losses. Secondly, eddy current and hysteresis losses should be also minimised. Eddy currents are internal currents that appear when there is a change in the magnetic field influencing the material. These internal currents lead to power losses. In contrast, hysteresis losses are caused by variations in the relationship between the magnetic field strength H and its flux density B. The latter two terms for losses are influenced by the frequency in a different way. In fact, eddy current losses increase roughly in proportion with frequency square whereas hysteresis losses increase only with the first power of the frequency. In addition, hysteresis losses can be minimised if the peak amplitude of the magnetic field intensity is far below the saturation region of the material. Thus, eddy current losses are expected to be determinant at high frequencies such as those used in EV wireless chargers. Both terms of power losses are also dependent on the temperature, so some refrigeration systems may be included for high-power applications in order to minimise these losses.

Considering these restrictions, ferrite can be considered a good component for complex coils structures in EV wireless chargers. It has high relative permeability ($\mu_r > 1000$) and a relatively low eddy current loss. The inclusion of ferrite cores in the coils leads to increased mutual inductance and some small variations in the self-inductance of the coil.

Once the material has been chosen, it is also necessary to decide its disposal and geometry. Ferrite materials are usually placed together with the coil. In particular, Manganese-Zinc ferrites are incorporated in EV wireless chargers. The dimensions and topologies of the ferrite components can be diverse, as illustrated in Fig. 3.12.

Bar ferrite tiles are commonly used in DD, DDQ and bipolar coils, as shown in Fig. 3.6.

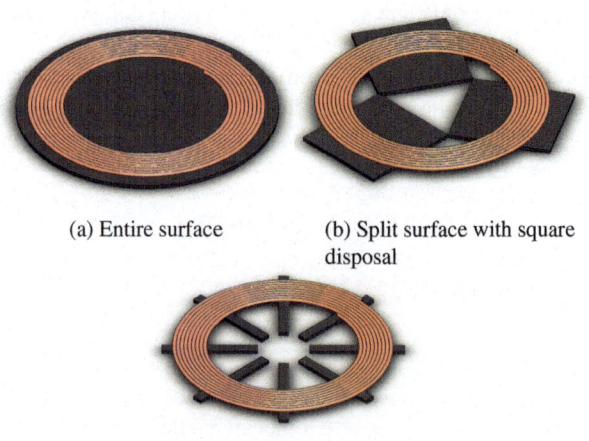

(a) Entire surface   (b) Split surface with square
                                 disposal

(c) Split surface with bar disposal

Fig. 3.12 Geometry of the ferrite plates

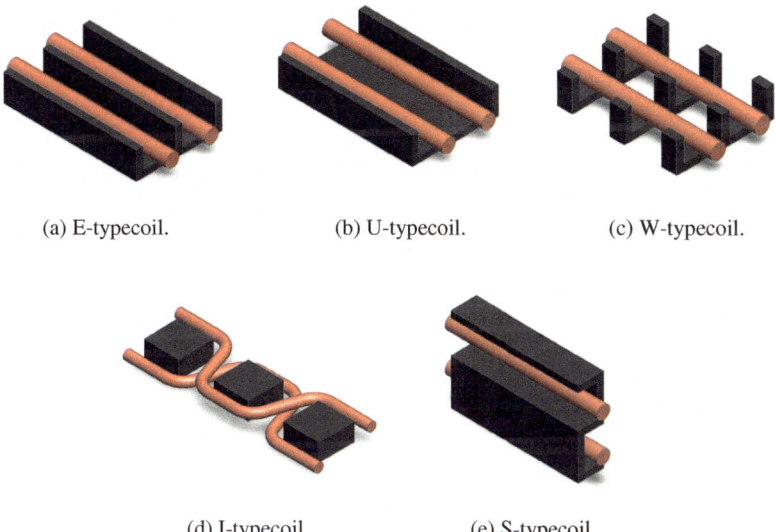

(a) E-typecoil.   (b) U-typecoil.   (c) W-typecoil.

(d) I-typecoil.   (e) S-typecoil.

Fig. 3.13 Ferrite core configurations for stretched coils

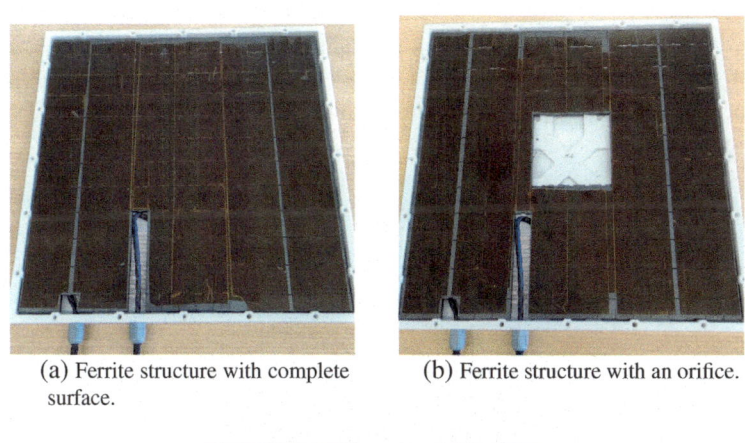

(a) Ferrite structure with complete   (b) Ferrite structure with an orifice.
surface.

(c) Ferrite structure with split disposal

Fig. 3.14 Ferrite geometries for laboratory coils [20, 21]

Ferrite core is also included in the stretched coils (those used on the primary side in some dynamic wireless charging) with different geometries. Figure 3.13 illustrates the most common configuration for this type of coil.

Despite these benefits, the inclusion of ferrite components complexifies the design as the equivalent electrical model of the charger is modified with some parameters that are not easily derived analytically. This includes the variation of self-inductance, mutual inductance and the equivalent coil resistance. Moreover, the variation of the mutual inductance of a pair of coils including ferrites with misalignment is not lineal. As a result, controlling an EV wireless charger with this type of coupler becomes more complex. Another important aspect to consider when analysing the suitability of this material is its weight. In some applications, such as UAVs, the weight of ferrite becomes a clear disadvantage for the receiver. Due to this inconvenience, Premo Group has developed a lighter material suitable for EV wireless chargers [4, 20, 21].

The images in Fig. 3.14 show the possible disposal of this polymer.

3.5   Magnetic Shielding

In order to comply with international recommendations on generated electromagnetic emissions, magnetic resonance chargers can incorporate some metallic plates underneath the coils. This technique is known as shielding. The effectiveness of shielding (SE_H) in a spatial point is defined as:

$$SE_H = 20 \log_{10}(H^i/H) \tag{3.25}$$

where H and H^i respectively correspond to the magnetic fields in the presence and in the absence of the shield at a point in the shielded region. H and H^i are also known as total and incident fields.

In the context of EV wireless chargers, the shielding techniques used can be classified into three main categories: conductive, active and reactive resonant. More than one of these techniques may be applied simultaneously. The techniques are described next.

3.5.1   Conductive Shielding

This method is supported by the Faraday-Lenz Law. The exposure of a conductive material (e.g. aluminium plate) to a time-varying magnetic field $B(t)$ induces currents on it. These currents are referred to as the Foucault $B'(t)$ currents. Due to the inductive component of the material, a new magnetic field is generated by the Foucault currents. $B'(t)$ has an opposite direction to $B(t)$, with the result being that the new magnetic field tries to cancel $B(t)$ on the opposite side of its incidence. In magnetic resonance chargers, the conductive material is usually a metal sheet placed underneath the

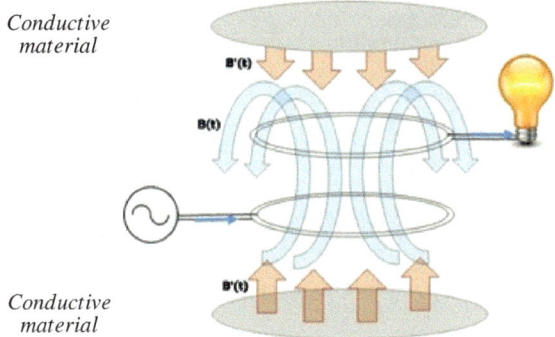

Fig. 3.15 Fundamental of conductive shielding in magnetic resonance chargers

coils. The physical behaviour of conductive shielding for this kind of application is illustrated in Fig. 3.15.

For conductive shields, SE_H can be approximated by [7]:

$$SE_H(\omega) = A(\omega) + R(\omega) + M(\omega) \tag{3.26}$$

where A corresponds to the absorption loss of the wave as it traverses the shield material, R represents the reflection loss due to the field reflection on the shield surface, and M models the additional effects of multiple reflections and transmissions. For magnetic resonance applications, the most significant term in Eq. 3.26 is the absorption loss. This term can be further described as:

$$A = 20 \log_{10}(e^{t/\delta}) \tag{3.27}$$

where t is the shield thickness and δ is the penetration depth.

When deciding the material and dimensions of the conductive shield, the penetration depth should be computed so that the thickness of the metal sheet should be at least the same as the penetration depth. This last term is defined as:

$$\delta = (\pi f \mu \sigma)^{-1/2} \tag{3.28}$$

where μ and σ are the permeability and the conductivity of the material respectively. Considering the particularities of copper and aluminium, they prove to be good candidates for conductive shieldings in EV wireless chargers. They have $\delta = 0.5$ mm at f = 20 kHz and $\delta = 0.2$ mm at f = 85 kHz [7].

The influence of the conductive shielding on the magnetic field is presented with the following example. In Fig. 3.16, a pair of coils is used for WPT with and without a conductive shield.

The resulting magnetic field is presented in Figs. 3.17 and 3.18 for both previous configurations at different phases of the primary current. These are simulations

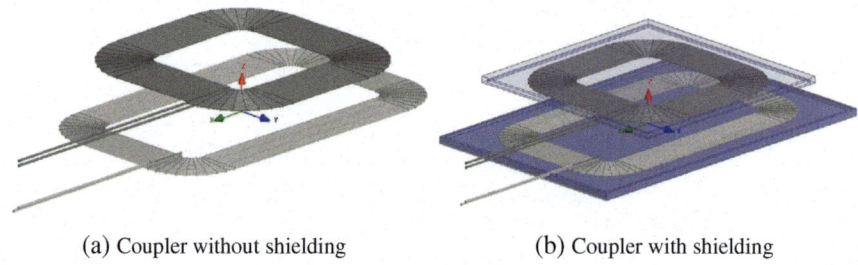

(a) Coupler without shielding   (b) Coupler with shielding

Fig. 3.16 Coils structure used for the electromagnetic emissions

obtained with Ansys Maxwell [1]. It can be observed that the shields prevent the magnetic field from traversing the area underneath them.

A conductive shield is usually applied with ferrite tiles to improve the magnetic field flux in the area comprised between the two coils. This is the usual construction of DD and DDQ coils. A generic configuration for both types of shielding is presented in Fig. 3.19 where the flux of the magnetic field $B(t)$ is altered to be maximised in the centre of the coils.

For the previous pair of coils, if we add a ferrite structure in the primary coil (Fig. 3.20), we will obtain the magnetic field in Fig. 3.21. This simulation reflects how the magnetic field is guided in the centre of the coils.

Although conductive shields help to mitigate the magnetic emissions outside the area comprised between the two coupled coils, its inclusion alters the electrical parameters of the coupler. Firstly, the conductive sheet behaves as an additional winding L_a with its associated losses R_a for the WPT system [16]. This additional coil is short-circuited. Secondly, the self-inductance of the coupler varies because of the new mutual inductances [11].

The effects of the conductive shielding on the equivalent circuit are illustrated in Fig. 3.22, where L_1 y L_2 are the primary and the secondary coil respectively. The additional mesh is caused by the shield and the coil in it is coupled with the primary and secondary coils as represented by the parameters M_{A1} and M_{A2}.

With circuital simplifications, it is possible to derive the next equivalent coupler in Fig. 3.23, where the self-inductances and the new mutual inductance of the equivalent coupler are lower than in the configuration of the original coils without the conductive shield [13]. In an opposite way, the equivalent resistances associated to the coils are increased.

Another serious problem of conductive shielding is the heat on the metallic sheet due to the eddy-current losses. If applied, the material of the coil should be selected to support this temperature.

Some software tools help to characterize the inductance and resistance of coils with complex geometries.

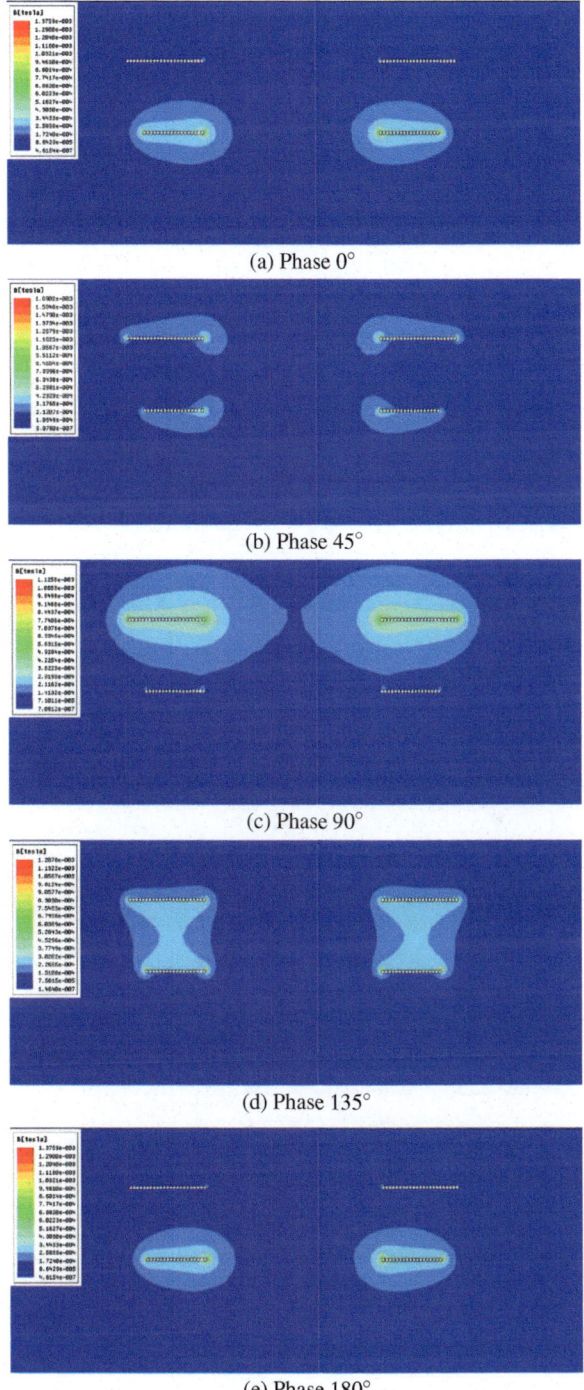

(a) Phase 0°

(b) Phase 45°

(c) Phase 90°

(d) Phase 135°

(e) Phase 180°

Fig. 3.17 Electromagnetic field of a coupler

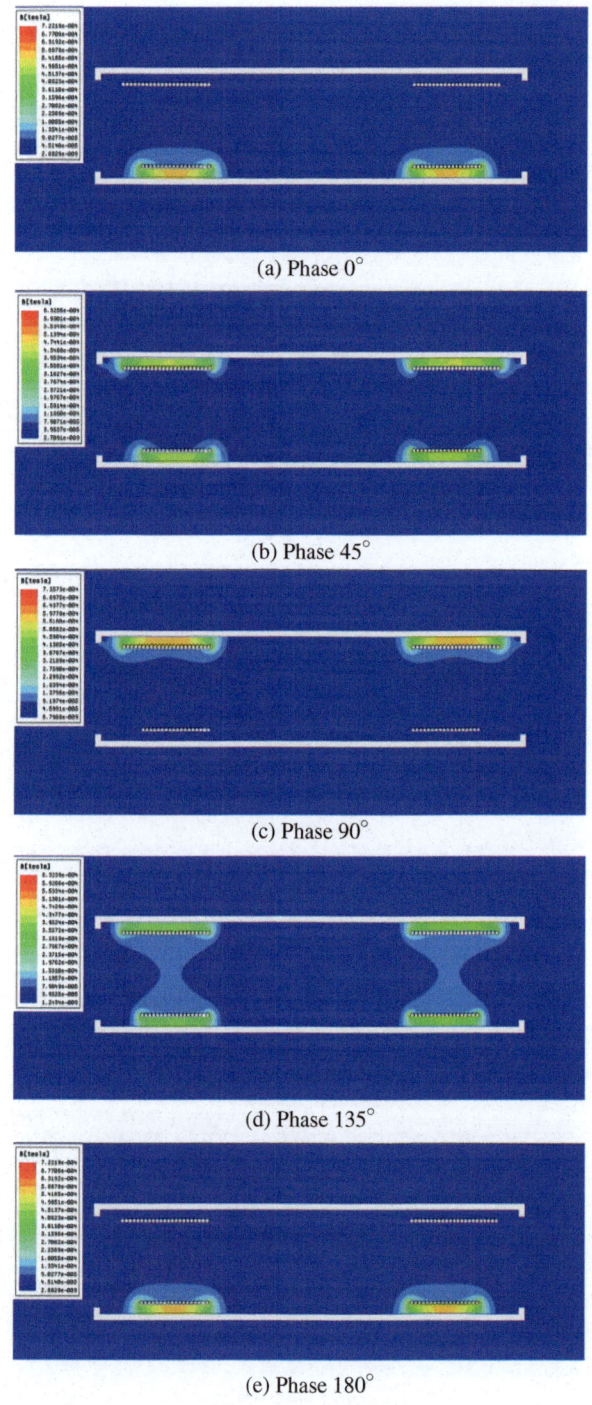

Fig. 3.18 Electromagnetic field of a coupler with shielding

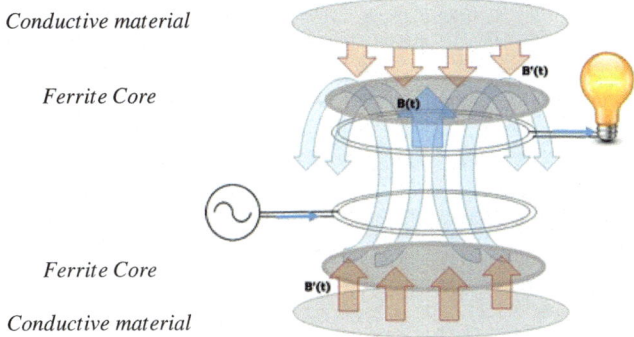

Conductive material

Ferrite Core

Ferrite Core

Conductive material

Fig. 3.19   Illustration of conductive shields combined with ferrite structures

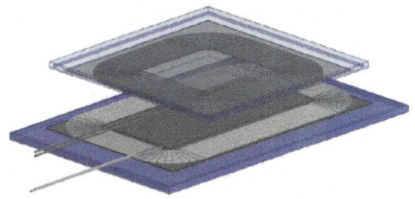

Fig. 3.20   Coupler with ferrite core

3.5.2   Active Shielding

This is based on external physical coils that are appropriately excited to generate a magnetic field in a direction opposite to the magnetic field originated by the coupler in those areas where it is of interest to mitigate the electromagnetic emissions. The main difficulty with this approach is deciding where to place these coils, their dimensions and the excitation current. A control scheme becomes necessary to adapt the power source of these additional sources to the instantaneous states of the magnetic field emissions. Additional power sources are also required, which increases the system costs and weight.

The most relevant implementations of this type of shielding were proposed for dynamic charging. In particular, they were applied in the PATH and the OLEV projects [8]. In the first project, the goal of the external coil was to cancel the magnetic field coming from the primary coil when no vehicle was detected for charging. The idea of this cancellation was to protect any pedestrians from exposure to magnetic fields. A different approach was used in the early generations of the OLEV project. Some additional capacitors were incorporated on the secondary side to compensate the secondary self-inductance. The value of the capacitor depended on the intensity of the magnetic field generated by the primary coil. A coordinated control was required.

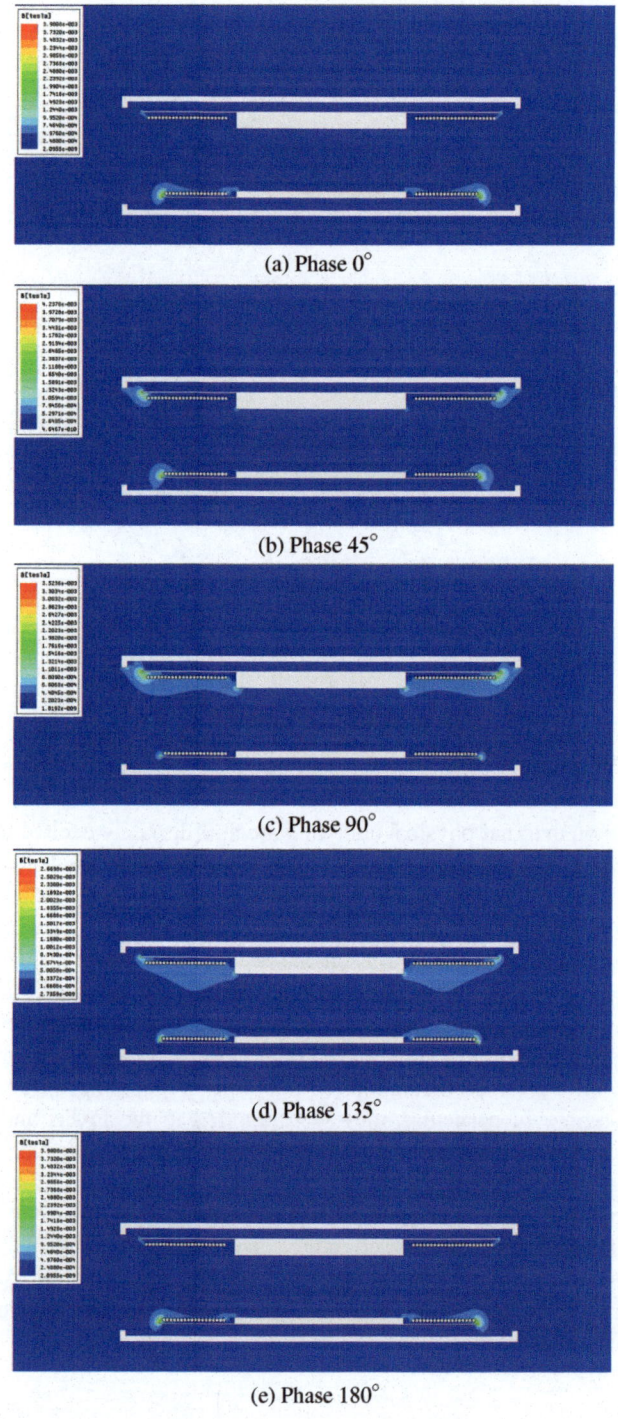

(a) Phase 0°

(b) Phase 45°

(c) Phase 90°

(d) Phase 135°

(e) Phase 180°

Fig. 3.21 Electromagnetic field of a coupler with ferrite and shielding

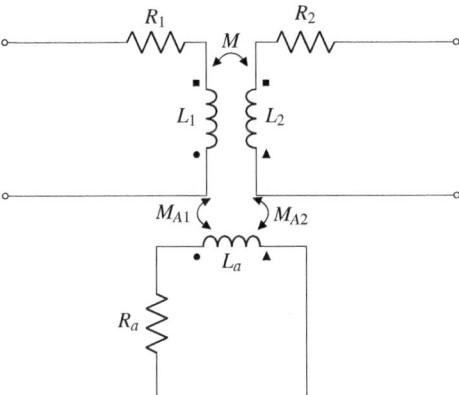

Fig. 3.22 Equivalent circuit of the coupler with conductive shielding

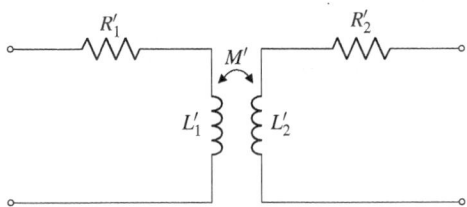

Fig. 3.23 Simplified equivalent circuit of a coupler with conductive shielding

3.5.3 Reactive Resonant Shielding

Reactive resonant shielding can be considered a particular feature of the active shield. Additional reactive components are incorporated into the WPT system but their excitation comes from the magnetic field naturally generated by the primary coil. The fifth generation of the OLEV project has developed three different implementations of reactive resonant shielding [9]. Specifically, extra capacitors and switch arrays to control the resonant frequency are the basis for this type of shielding in the OLEV project.

References

1. ANSYS Maxwell: Low Frequency Electromagnetic Field Simulation. https://www.ansys.com/products/electronics/ansys-maxwell
2. COMSOL Multiphysics® Modeling Software. https://www.comsol.es/
3. JMAG: Simulation Technology for Electromechanical Design. https://www.jmag-international.com/
4. PREMO Group - Grupo Premo. https://www.grupopremo.com/

5. Boys, J.T., Covic, G.A.: The inductive power transfer story at the University of Auckland. IEEE Circuits Syst. Mag. **15**(2), 6–27 (2015). https://doi.org/10.1109/MCAS.2015.2418972, http://ieeexplore.ieee.org/document/7110451/
6. Budhia, M., Boys, J.T., Covic, G.A., Huang, C.Y.: Development of a single-sided flux magnetic coupler for electric vehicle IPT charging systems. IEEE Trans. Ind. Electron. **60**(1), 318–328 (2013). https://doi.org/10.1109/TIE.2011.2179274, http://ieeexplore.ieee.org/document/6099605/
7. Campi, T., Cruciani, S., Feliziani, M.: Magnetic shielding of wireless power transfer systems. In: International Symposium on Electromagnetic Compatibility, vol. 10, pp. 422–425. Tokyo (2014)
8. Choi, S.Y., Gu, B.W., Jeong, S.Y., Rim, C.T.: Advances in wireless power transfer systems for roadway-powered electric vehicles. IEEE J. Emerg. Sel. Top. Power Electron. **3**(1), 18–36 (2015). https://doi.org/10.1109/JESTPE.2014.2343674, http://ieeexplore.ieee.org/document/6871280/
9. Choi, S.Y., Gu, B.W., Lee, S.W., Lee, W.Y., Huh, J., Rim, C.T.: Generalized active EMF cancel methods for wireless electric vehicles. IEEE Trans. Power Electron. **29**(11), 5770–5783 (2014). https://doi.org/10.1109/TPEL.2013.2295094, http://ieeexplore.ieee.org/document/6684288/
10. Fotopoulou, K., Flynn, B.W.: Wireless power transfer in loosely coupled links: coil misalignment model. IEEE Trans. Magn. **47**(2), 416–430 (2011). https://doi.org/10.1109/TMAG.2010.2093534, http://ieeexplore.ieee.org/document/5639082/
11. Geselowtiz, D., Hoang, Q., Gaumond, R.: The effects of metals on a transcutaneous energy transmission system. IEEE Trans. Biomed. Eng. **39**(9), 928–934 (1992). https://doi.org/10.1109/10.256426, http://ieeexplore.ieee.org/document/256426/
12. International Commission on Non-Ionizing Radiation Protection: Guidelines for limiting exposure to time-varying electric and magnetic fields (1 Hz to 100 kHz). Health Phys. **99**(6), 818–836 (2010). https://doi.org/10.1097/HP.0b013e3181f06c86, http://www.ncbi.nlm.nih.gov/pubmed/21068601
13. Kim, J., Kim, J., Kong, S., Kim, H., Suh, I.S., Suh, N.P., Cho, D.H., Kim, J., Ahn, S.: Coil design and shielding methods for a magnetic resonant wireless power transfer system. Proc. IEEE **101**(6), 1332–1342 (2013). https://doi.org/10.1109/JPROC.2013.2247551, http://ieeexplore.ieee.org/document/6480778/
14. Luo, Z., Wei, X.: Mutual inductance analysis of planar coils with misalignment for wireless power transfer systems in electric vehicle. In: 2016 IEEE Vehicle Power and Propulsion Conference (VPPC), pp. 1–6. IEEE (2016). https://doi.org/10.1109/VPPC.2016.7791733, http://ieeexplore.ieee.org/document/7791733/
15. Luo, Z., Wei, X.: Analysis of square and circular planar spiral coils in wireless power transfer system for electric vehicles. IEEE Trans. Ind. Electron. **65**(1), 331–341 (2018). https://doi.org/10.1109/TIE.2017.2723867, http://ieeexplore.ieee.org/document/7968494/
16. Mauro Feliziani, S.C.: Mitigation of the magnetic field generated by a wireless power transfer (WPT) system without reducing the WPT efficiency. In: International Symposium on Electromagnetic Compatibility (2013). https://ieeexplore.ieee.org/document/6653375
17. Nagendra, G.R., Covic, G.A., Boys, J.T.: Determining the physical size of inductive couplers for IPT EV systems. IEEE J. Emerg. Sel. Top. Power Electron. **2**(3), 571–583 (2014). https://doi.org/10.1109/JESTPE.2014.2302295, http://ieeexplore.ieee.org/document/6720158/
18. New England Wire Technologies: Litz wire: technical information (2009). http://www.litzwire.com/nepdfs/Litz_Technical.pdf
19. Patil, D., McDonough, M.K., Miller, J.M., Fahimi, B., Balsara, P.T.: Wireless power transfer for vehicular applications: overview and challenges. IEEE Trans. Transp. Electrific. **4**(1), 3–37 (2018). https://doi.org/10.1109/TTE.2017.2780627, https://ieeexplore.ieee.org/document/8168345/
20. Rojas Cuevas, A., Navarro Pérez, F.E., Cañete Cabeza, C.: Device and method for winding a flexible elongated inductor (2016)
21. Rojas Cuevas, A., Navarro Pérez, F.E., Cañete Cabeza, C.: Elongated flexible inductor and elongated flexible low frequency antenna (2016)

22. Rosskopf, A., Bar, E., Joffe, C.: Influence of inner skin- and proximity effects on conduction in litz wires. IEEE Trans. Power Electron. **29**(10), 5454–5461 (2014). https://doi.org/10.1109/TPEL.2013.2293847, http://ieeexplore.ieee.org/document/6678811/
23. Shkolnikov, Y.P., Bailey, W.H.: Electromagnetic interference and exposure from household wireless networks. In: 2011 IEEE Symposium on Product Compliance Engineering Proceedings, pp. 1–5. IEEE (2011). https://doi.org/10.1109/PSES.2011.6088244, http://ieeexplore.ieee.org/document/6088244/
24. Triviño, A., Gonzalez-Gonzalez, J.M., Aguado, J.A.: Theoretical analysis of the efficiency of a V2G wireless charger for electric vehicles. Trans. Environ. Electr. Eng. **3**(1), 9 (2018). https://doi.org/10.22149/teee.v3i1.118, https://teee.eu/index.php/TEEE/article/view/118
25. Triviño-Cabrera, A., Aguado, J., González, J.: Analytical characterisation of magnetic field generated by ICPT wireless charger. Electron. Lett. **53**(13), 871–873 (2017). https://doi.org/10.1049/el.2017.0968, http://digital-library.theiet.org/content/journals/10.1049/el.2017.0968
26. Wojda, R.P., Kazimierczuk, M.K.: Winding resistance and power loss of inductors with litz and solid-round wires. IEEE Trans. Ind. Appl. **54**(4), 3548–3557 (2018). https://doi.org/10.1109/TIA.2018.2821647, https://ieeexplore.ieee.org/document/8329131/
27. Zhang, W., Zhang, T., Guo, Q., Shao, L., Zhang, N., Jin, X., Yang, J.: High-efficiency wireless power transfer system for 3D, unstationary free-positioning and multi-object charging. IET Electr. Power Appl. **12**(5), 658–665 (2018). https://doi.org/10.1049/iet-epa.2017.0581, https://digital-library.theiet.org/content/journals/10.1049/iet-epa.2017.0581
28. Zhang, Z., Pang, H., Lee, C.H.T., Xu, X., Wei, X., Wang, J.: Comparative analysis and optimization of dynamic charging coils for roadway-powered electric vehicles. IEEE Trans. Magn. **53**(11), 1–6 (2017). https://doi.org/10.1109/TMAG.2017.2736164, http://ieeexplore.ieee.org/document/8002591/

Chapter 4
Compensation Networks

4.1 Introduction

In contrast to pure inductive chargers, magnetic resonance technology uses reactive structures in the primary and secondary coils in order to make the system operate under resonant conditions. Due to the dimensions of the coils, the parasitic capacitance is not sufficient to ensure the resonance in the operational frequency range. Note that certain applications restrict the frequency at which the system works. This is the case with Qi-systems, RFID products, biomedical applications and even EV wireless chargers that are compliant with SAE J2954 specifications. Consequently, additional reactive structures, known as the compensation networks, are incorporated into the systems to adjust the operational frequency.

Although a single compensation network on the primary side is theoretically able to force the system to operate on resonance, in high-power applications it is more convenient to rely on two compensation networks to increase the control options in the systems. Thus, there is a compensation network connected to the primary coil and another to the secondary coil. Figure 4.1 illustrates the generic diagram of a magnetic resonance wireless charger.

The main functionality of the compensation network is to resonate with its associated coil so that the reactive power supply is minimized. In the context of wireless chargers for EVs, the compensation networks are designed to improve both the efficiency and the power transfer capability.

To understand the importance of the compensation networks, the analysis of the coupled coils without these elements is presented below. The equivalent circuit of this inductive charger is illustrated in Fig. 4.2. The voltage source models the first harmonic of the signal generated by the power converters. It has a rms value equal to V_1 and its angular frequency is ω.

The system can be described with the following two equations:

$$\mathbf{V}_1 = (R_1 + j\omega L_1)\,\mathbf{I}_1 - j\omega M \mathbf{I}_2 \tag{4.1}$$

© Springer Nature Switzerland AG 2020
A. Triviño-Cabrera et al., *Wireless Power Transfer for Electric Vehicles: Foundations and Design Approach*, Power Systems,
https://doi.org/10.1007/978-3-030-26706-3_4

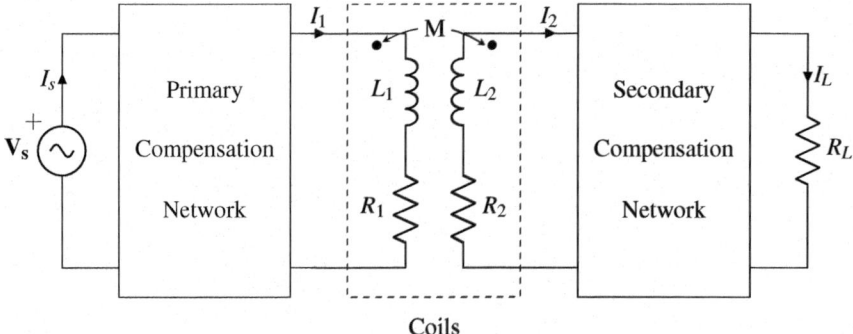

Fig. 4.1 Generic diagram for magnetic resonance wireless chargers with compensation networks

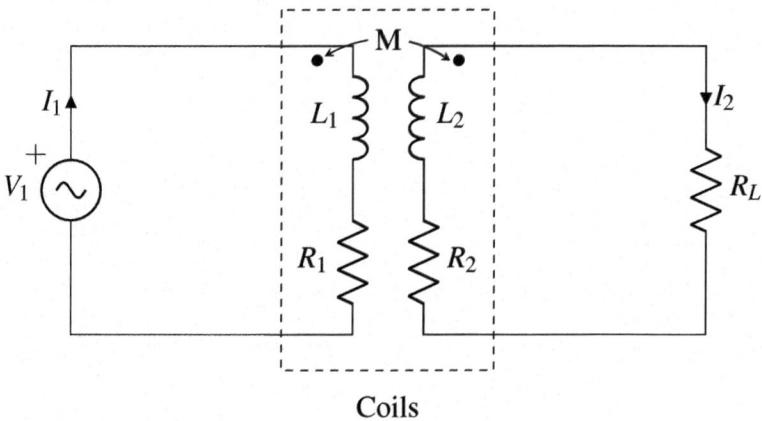

Fig. 4.2 Equivalent circuit of an inductive charger without any compensation network

$$j\omega M \mathbf{I}_1 = (R_2 + j\omega L_2 + R_L)\mathbf{I}_2 \qquad (4.2)$$

From the last equation, we can infer that:

$$\frac{\mathbf{I}_1}{\mathbf{I}_2} = \frac{j\omega L_2 + R_2 + R_L}{j\omega M} \qquad (4.3)$$

The efficiency (η) of the system is defined as the ratio of the power delivered to the load between the load generated by the source. Formally, this is expressed as:

$$\eta = \frac{R_L I_2^2}{R_L I_2^2 + R_2 I_2^2 + R_1 I_1^2} = \frac{R_L}{R_L + R_2 + R_1 (\frac{I_1}{I_2})^2} \qquad (4.4)$$

Without the secondary compensation network, we use the relationship between the currents expressed in Eq. 4.3 so that:

$$\eta = \frac{R_L}{R_L + R_2 + R_1 \frac{(R_2+R_L)^2+(\omega L_2)^2}{\omega^2 M^2}} \qquad (4.5)$$

We can observe that when the compensation network is not included, the denominator of the previous definition has a term due to the self inductance L_2. This term is always positive so that the efficiency is lower than in an alternative configuration where this term is suppressed. A correctly designed compensation network in the secondary side is able to eliminate this term and, in turn, increase the efficiency.

The primary compensation network is included in the system to compensate the reactive power generated by the source. As a consequence, the power delivered to the load is increased. However, the application of the complete theorem of the maximum power transfer is not performed as this would imply 50% efficiency.

The reactive components of the compensation network can be arranged with different topologies leading to other benefits in some specific configurations. As shown in [30], the most significant additional functionalities are:

- Reduce the need for an additional control system. Some compensation topologies make the system operate as CV or CC output. This implies that the wireless charger can be highly tolerant to changes in the electrical load, which commonly occurs during the charge process. As a result, the electronics of the control system in the wireless charger can be avoided or simplified.
- Minimize the losses of the power converters. As previously explained, the compensation networks are designed so that the reactance of the input impedance of the source is null. For the power converters, this means that Zero-Phase Angle shifting is usually set. This setting reduces the losses associated with the power converters and, in turn, the efficiency of the whole system (including that of the power converters) is improved [4].

The most simple topologies for the compensation networks opt for just one capacitor in these structures. They are referred to as the mono-resonant compensation topology, because the resonant tank (the capacitor with its corresponding coil) resonates at only one frequency. Alternatively, there are other compensation topologies which rely on more reactive elements. They are known as multi-resonant compensation topologies.

When designing an EV wireless charger, one of the decisions that must be taken is which compensation network should be used and how it should be configured. When making the choice, it is necessary to understand the properties of the compensation topologies. In particular, it is essential to know how stable the system is when changes in the operational frequency occur. As a generic approach, the bifurcation phenomenon is avoided.

This fourth chapter firstly analyses the bifurcation phenomenon. Next, the most common compensation topologies in EV wireless chargers are described in depth from an electrical point of view. For the study performed in this chapter, the first harmonic approximation is used so that the output voltage of the primary converters is modelled as a pure sinusoidal wave.

4.2 Stability of the Compensation Networks

Magnetic resonance chargers are usually designed to operate under resonant conditions. However, their real operation may occur with a different configuration. For instance, the wireless charger could be designed to work with a specific gap between the coils but it can also operate with a different separation between the two inductive elements. As a result, the mutual inductance and the reflected impedance vary. Thus, this deviation could mean that the system does not operate under resonant conditions. To cope with this eventuality, some control algorithms adjust the operational frequency of the resonant system [10, 15, 17]. This strategy is only valid if the system is stable when the frequency is modified.

To know if this condition holds, the bifurcation phenomenon should be analysed and taken into account in the design process. In the WPT context, the bifurcation phenomenon refers to the situation in which there is more than one (Zero Phase Angle (ZPA) frequency. As a result, some electrical magnitudes (e.g power delivered to the load) have more than one maximum at the frequencies with null reactance. If this effect has not been evaluated, the peak values could damage or deteriorate some of the electronics in the wireless charger.

As stated in [30], the bifurcation depends on the loading condition, the compensation topologies and the values of the reactive components. The work in [29] addresses how to define the coils and their corresponding capacitors to avoid the bifurcation phenomenon in the four simple compensation topologies. Some economical aspects could also be of relevance for this setting as presented in [12].

A phenomenon associated with the bifurcation is frequency splitting [21]. Frequency splitting relates the number of ZPA frequencies to the coupling coefficient. Specifically, a critical coupling coefficient is defined for a configuration of two identical coupled coils. As presented in [26], when the coupling coefficient exceeds the critical one, the bifurcation phenomenon manifests itself. The critical coupling coefficient is usually close to 1, which is greater than the common coupling coefficient of wireless chargers in EV. This is why the effects of frequency splitting are not commonly evaluated in this kind of application. Nevertheless, frequency splitting is studied in depth in mid-range WPT systems.

The description of the main compensation topologies in the following sections will also address the evaluation of their stability in relation to the bifurcation phenomenon.

4.3 Mono-resonant Compensation Networks

There are four mono-resonant compensation topologies. They are commonly identified with two terms separated by a dash. The first term represents the connection (series or in parallel) between the capacitor and the primary coil. Similarly, the second term reflects the connection between the compensation network in the secondary side and its corresponding coil. Thus, the mono-resonant compensa-

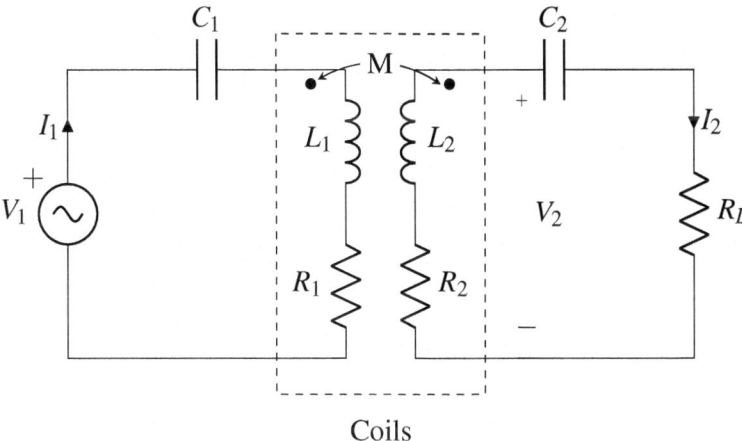

Fig. 4.3 Equivalent circuit of a magnetic resonance wireless charger with a Series–Series compensation topology

tion topologies are: Series–Series (SS), Series–Parallel (SP), Parallel–Series (PS) or Parallel–Parallel (PP).

The next sections analyse the compensation topology and describe how they are designed. On that subject, it is worth noting that wireless chargers for medium and high power are designed to maximize the efficiency. An evaluation of when the bifurcation occurs is also included.

4.3.1 Series–Series Topology

The Series–Series compensation topology is illustrated in the Fig. 4.3:

With a mesh-based analysis, we can state that:

$$V_1 = \left(R_1 + j\omega L_1 + \frac{1}{j\omega C_1} \right) I_1 - j\omega M I_2 \tag{4.6}$$

$$V_2 = [(R_2 + j\omega L_2) I_2 - j\omega M I_1] = \left(R_L + \frac{1}{j\omega C_2} \right) I_2 \tag{4.7}$$

From the two previous equations, we can infer the input impedance Z_{in} as follows:

$$Z_{in} = \frac{V_1}{I_1} = R_1 + j\left(\omega L_1 - \frac{1}{\omega C_1} \right) + \frac{\omega^2 M^2}{(R_2 + R_L) + j\left(\omega L_2 - \frac{1}{\omega C_2} \right)} \tag{4.8}$$

In order to have a purely ohmic input impedance, the reactive term must be null. This condition is expressed with the following equation:

$$\omega L_1 - \frac{1}{\omega C_1} = -Imag\left(\frac{\omega^2 M^2}{(R_2 + R_L) + j\left(\omega L_2 - \frac{1}{\omega C_2}\right)}\right) \tag{4.9}$$

After developing this further, we obtain:

$$\omega L_1 - \frac{1}{\omega C_1} = \frac{\omega^2 M^2\left(\omega L_2 - \frac{1}{\omega C_2}\right)}{(R_2 + R_L)^2 + j\left(\omega L_2 - \frac{1}{\omega C_2}\right)^2} \tag{4.10}$$

This last equation is verified when the two pairs of coils and their corresponding capacitor constitute two independent resonant tanks at the same operational frequency (ω_0).

$$\omega_0 = \frac{1}{\sqrt{L_1 C_1}} = \frac{1}{\sqrt{L_2 C_2}} \tag{4.11}$$

Thus, to design a SS compensation network, the capacitors can be defined with their corresponding coils once the frequency is set according to the application specifications.

The system efficiency is defined as the ratio between the active power delivered to the load and the active power generated by the source. This parameter can be formally expressed as:

$$\eta = \frac{R_L I_2^2}{R_1 I_1^2 + R_2 I_2^2 + R_L I_2^2} = \frac{R_L}{\frac{R_1 I_1^2}{I_2^2} + R_2 + R_L} \tag{4.12}$$

Please note that the inclusion of the compensation network has led to the suppression of the term due to the self-inductance L_2 in Eq. 4.3, which defined a purely inductive system.

From Eq. 4.7, we can infer the relationship between the modules of the I_1 and I_2 currents.

$$\left|\frac{I_1}{I_2}\right| = \frac{R_2 + R_L}{\omega_0 M} \tag{4.13}$$

So that the system efficiency is equal to:

$$\eta = \frac{R_L}{R_1\left(\frac{R_2 + R_L}{\omega_0 M}\right)^2 + R_2 + R_L} \tag{4.14}$$

In order to maximize the efficiency, the denominator of the previous expression should be minimized. This implies that:

$$\frac{R_2 + R_L}{\omega_0 M} \ll 1 \tag{4.15}$$

Thus,

$$\omega_0 \gg \frac{R_2 + R_L}{M} \tag{4.16}$$

If the relationship in Eq. 4.16 holds, the efficiency corresponds to:

$$\eta \cong \frac{R_L}{R_2 + R_L} \tag{4.17}$$

Based on Eq. 4.9, the work in [29] deduces that the bifurcation phenomenon is avoided when the following condition is verified:

$$q_1 > \frac{4q_2^3}{4q_2^3 - 1} \tag{4.18}$$

where q_1 and q_2 are the quality factors of the primary and secondary coils respectively.

The next set of graphs in Fig. 4.4 illustrates the behaviour of two magnetic resonance systems. The one associated with the black curves was designed avoiding the bifurcation phenomenon. In contrast, the results with the blue lines correspond to a magnetic resonance charger where the condition in Eq. 4.18 is not true. Specifically, the parameters of the two systems are summarized in Table 4.1.

The bifurcation analysis presented in Fig. 4.4 (dashed line) shows an increase in the electrical magnitudes of the system when working with frequency deviations. Although all of them show this increment, the most pronounced is in the primary side, where current and capacitor voltage reach up to 250% of the nominal values. Despite operating with bifurcation, efficiency remains quite high even with large variations in frequency.

However, if the design of the compensation system is conducted in an appropriate manner to avoid the bifurcation phenomenon (solid line), the electrical magnitudes are maintained at values similar to the nominal values for variations in the range of ±5% of the nominal frequency value. For frequency deviations greater than this value, a more pronounced decrease in system efficiency can be observed with respect to bifurcated designs.

Bearing in mind the previous analysis, we observe for the SS compensation topology that:

- The adjustment of the primary and secondary compensation networks can be done independently as expressed by Eq. 4.13.
- It makes the system operates as a constant voltage source.
- High tolerance to misalignment. According to Eq. 4.11, the imaginary part of the input impedance does not depend on the mutual inductance when the two tanks are adjusted to resonate at the operational frequency.
- It usually leads to reduced weights of the coils [24].

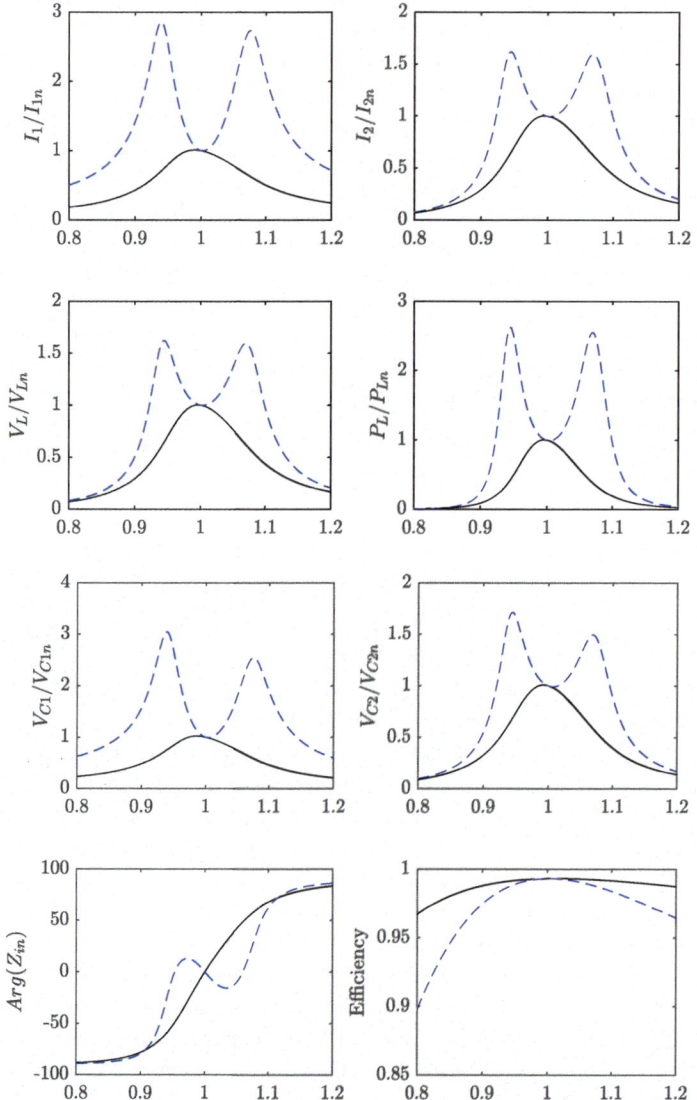

Fig. 4.4 Variation of the main electrical magnitudes for two magnetic resonance systems based on the SS compensation topology. Stable configuration in black and the system with bifurcation in blue

Table 4.1 Parameters used in the stability analysis with SS compensation network

Component	Non-bifurcation value (μH)	Bifurcation value (μH)
L_1	475	165
L_2	152	409
M	37	36

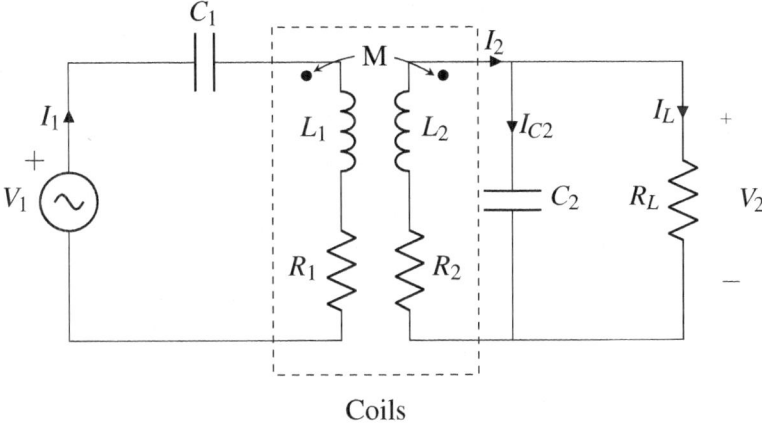

Coils

Fig. 4.5 Equivalent circuit of a magnetic resonance wireless charger with a Series–Parallel compensation topology

- Due to its symmetry, SS compensation topology is a common option in bidirectional wireless chargers. The equivalent structures on the primary and secondary sides eases the development of similar control schemes on both the primary and the secondary sides.

4.3.2 Series–Parallel Topology

The Fig. 4.5 reflects the equivalent circuit of a magnetic resonance charger with Series–Parallel compensation networks:

A mesh-based analysis results in:

$$\mathbf{V_1} = \left(R_1 + j\omega L_1 + \frac{1}{j\omega C_1} \right) \mathbf{I_1} - j\omega M \mathbf{I_2} \tag{4.19}$$

$$\mathbf{V_2} = -\left(R_2 + j\omega L_2 + \right) \mathbf{I_2} + j\omega M \mathbf{I_1} = R_L \mathbf{I_L} = \frac{\mathbf{I_{C2}}}{j\omega C_2} \tag{4.20}$$

According to Kirchhoff's Current Law, we can state that:

$$\mathbf{I_2} = \mathbf{I_{C2}} + \mathbf{I_L} \tag{4.21}$$

As the capacitor C_2 is in parallel with the load resistance, we conclude that:

$$\mathbf{I_{C2}} = j\omega R_L C_2 \mathbf{I_L} \tag{4.22}$$

Reformulating Eqs. 4.19 and 4.20 as a function of currents $\mathbf{I_1}$ and $\mathbf{I_L}$, we obtain:

$$\mathbf{V_1} = \left(R_1 + j\omega L_1 + \frac{1}{j\omega C_1} \right) \mathbf{I_1} - j\omega M \left(j\omega C_2 R_L + 1 \right) \mathbf{I_L} \qquad (4.23)$$

$$0 = \left(R_2 + j\omega L_2 \right) \left(j\omega C_2 R_L + 1 \right) \mathbf{I_L} - j\omega M \mathbf{I_1} + R_L \mathbf{I_L} \qquad (4.24)$$

From this last Equation, we derive that:

$$\mathbf{I_L} = \frac{j\omega M}{R_L + \left(R_2 + j\omega L_2 \right) \left(j\omega C_2 R_L + 1 \right)} \mathbf{I_1} \qquad (4.25)$$

Replacing $\mathbf{I_L}$ in 4.23, we define the input impedance (Z_{in}) as:

$$Z_{in} = \frac{\mathbf{V_1}}{\mathbf{I_1}} = \left(R_1 + j\omega L_1 + \frac{1}{j\omega C_1} \right) + \frac{\omega^2 M^2 \left(j\omega C_2 R_L + 1 \right)}{R_L + \left(R_2 + j\omega L_2 \right) \left(j\omega C_2 R_L + 1 \right)} \qquad (4.26)$$

To avoid the reactive power generated by the source, we have to impose that:

$$\omega L_1 - \frac{1}{\omega C_1} = -Imag \left(\frac{\omega^2 M^2 \left(j\omega C_2 R_L + 1 \right)}{R_L + \left(R_2 + j\omega L_2 \right) \left(j\omega C_2 R_L + 1 \right)} \right) \qquad (4.27)$$

To proceed to eliminate the reactance expressed in Eq. 4.27, the first step is to make the secondary tank resonate at a specified angular frequency (ω_0).

$$\omega_0 = \frac{1}{\sqrt{L_2 C_2}} \qquad (4.28)$$

Then, if we consider that R_2 is negligible in comparison with R_L, we can simplify Eq. 4.27 as:

$$\omega_0 L_1 - \frac{1}{\omega_0 C_1} \approx \frac{\omega_0 M^2}{L_2} \qquad (4.29)$$

From this last expression, we can conclude that the proper value for the capacitor C_1 in the primary resonant network is:

$$C_1 = \frac{L_2^2 C_2}{L_1 L_2 - M^2} \qquad (4.30)$$

In addition, Eq. 4.27 helps us to analyse under which circumstances there is one or multiple roots for ω, that is, when bifurcation occurs. According to [29], the SP compensation topology is stable if:

$$q_1 > q_2 + \frac{1}{q_2} \tag{4.31}$$

For the same configurations of Table 4.2, the electrical parameters exhibit the relationship presented in Fig. 4.6 with the variations of the operational frequency.

The bifurcation analysis of the SP topology shows a similar performance to the SS topology, as it presents an increment in electrical magnitudes when the frequency is altered from its nominal value. Nevertheless, this variation is lower than with SS topology and the bifurcation phenomena is not as critical. The efficiency also declines less than SS topology with frequency variations both with bifurcation and without this phenomenon and, with large deviations, the latter displays greater efficiency than the former.

4.3.3 Parallel–Series Topology

The Parallel–Series (PS) compensation topology is defined with the equivalent circuit in Fig. 4.7.

As can be observed, the primary coil (including its internal resistance R_1) is in parallel with the voltage source so:

$$\mathbf{V_1} = (R_1 + j\omega L_1)\,\mathbf{I_1} - j\omega M \mathbf{I_2} \tag{4.32}$$

The source voltage is also in parallel with the capacitor C_1 so:

$$\mathbf{I_{C1}} = j\omega C_1 \mathbf{V_1} \tag{4.33}$$

Alternatively, the analysis of the mesh in the secondary side determines that:

$$\mathbf{V_2} = [(R_2 + j\omega L_2)\,\mathbf{I_2} - j\omega M \mathbf{I_1}] = \left(R_L + \frac{1}{j\omega C_2} \right) \mathbf{I_2} \tag{4.34}$$

From this last equation, we can formulate that:

$$\mathbf{I_2} = \frac{j\omega M}{(R_2 + R_L) + j\left(\omega L_2 - \frac{1}{\omega C_2} \right)} \mathbf{I_1} \tag{4.35}$$

To derive the input impedance Z_{in}, we first establish the relationship between the source voltage V_1 and the current I_1 as follows:

$$\mathbf{V_1} = \left[(R_1 + j\omega L_1) + \frac{\omega^2 M^2}{(R_2 + R_L) + j\left(\omega L_2 - \frac{1}{\omega C_2} \right)} \right] \mathbf{I_1} \tag{4.36}$$

Table 4.2 Parameters used in the stability analysis with SP compensation network

Property	Non-bifurcation values (μH)	Bifurcation values (μH)
L_1	165	473
L_2	11.5	4.3
M	6	6

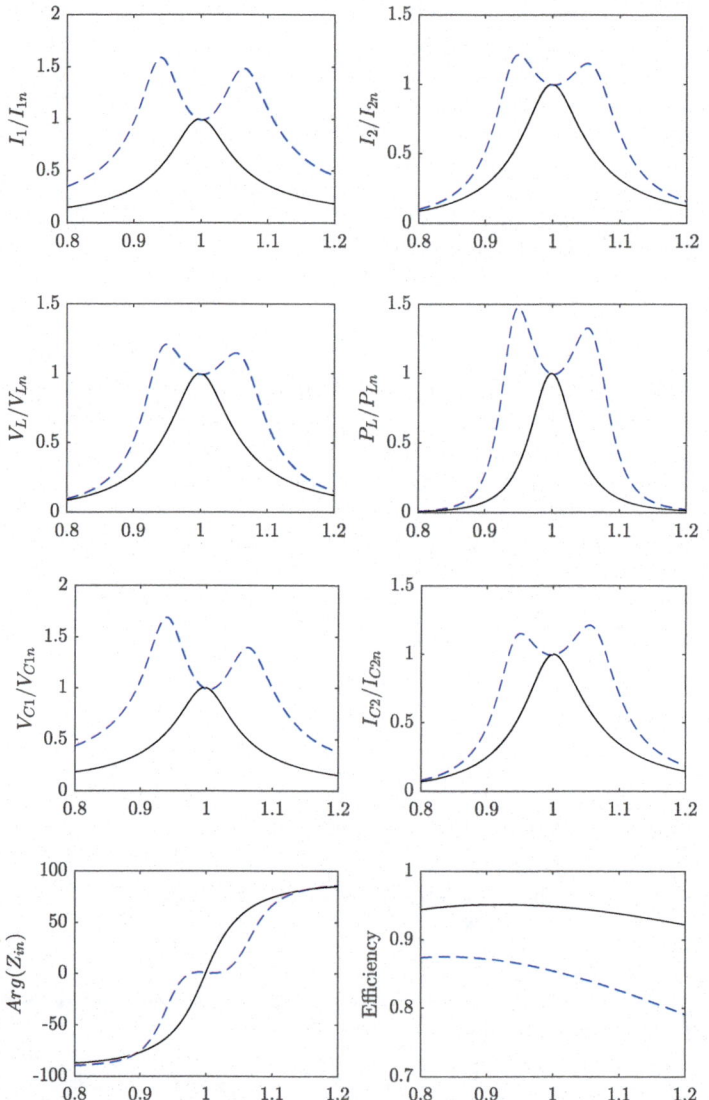

Fig. 4.6 Variation of the main electrical magnitudes for two magnetic resonance systems based on the SP compensation topology. Stable configuration is shown in black and the system with bifurcation is shown in blue

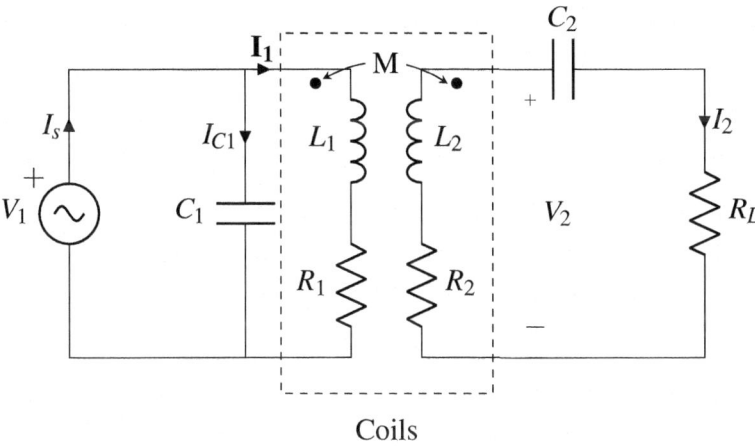

Fig. 4.7 Equivalent circuit of a magnetic resonance wireless charger with a Parallel–Series compensation topology

Applying Kirchhoff's Current Law, we can establish that $\mathbf{I_s} = \mathbf{I_{C1}} + \mathbf{I_1}$, so that:

$$\mathbf{V_1} = \left[(R_1 + j\omega L_1) + \frac{\omega^2 M^2}{(R_2 + R_L) + j\left(\omega L_2 - \frac{1}{\omega C_2}\right)} \right] (\mathbf{I_s} - j\omega C_1 \mathbf{V_1}) \quad (4.37)$$

As a result, the input impedance Z_{in} is defined as:

$$Z_{in} = \frac{\mathbf{V_1}}{\mathbf{I_s}} = \frac{(R_1 + j\omega L_1) + \frac{\omega^2 M^2}{(R_2+R_L)+j\left(\omega L_2-\frac{1}{\omega C_2}\right)}}{1 + j\omega C_1 (R_1 + j\omega L_1) + \frac{\omega^2 M^2}{(R_2+R_L)+j\left(\omega L_2-\frac{1}{\omega C_2}\right)}} \quad (4.38)$$

If we impose that the secondary compensation network resonates at the operational angular frequency (ω_0), then:

$$\omega_0 = \frac{1}{\sqrt{L_2 C_2}} \quad (4.39)$$

Under this assumption, the impedance Z_{in} is simplified as:

$$Z_{in} = \frac{\mathbf{V_1}}{\mathbf{I_s}} = \frac{(R_1 + j\omega_0 L_1) + \frac{\omega_0^2 M^2}{R_2+R_L}}{1 + j\omega_0 C_1 (R_1 + j\omega_0 L_1) + \frac{\omega_0^2 M^2}{R_2+R_L}} \quad (4.40)$$

As in the compensation topologies analysed previously, it is advisable to eliminate the reactive power generated by the source. This suppression is achieved if the input impedance has a null imaginary part. The reactance of the input impedance is:

$$Imag(Z_{in}) = \frac{\omega_0 L_1 \left(1 - \omega_0^2 L_1 C_1\right) - \omega_0 C_1 \left(R_1 + \frac{\omega_0^2 M^2}{R_2 + R_L}\right)^2}{\left(1 - \omega_0^2 L_1 C_1\right)^2 + \omega_0^2 C_1^2 \left(R_1 + \frac{\omega_0^2 M^2}{R_2 + R_L}\right)^2} \qquad (4.41)$$

Considering that R_1 and R_2 are much lower than R_L, the reactance is null when the value of C_1 is computed as follows:

$$C_1 = \frac{L_2 C_2}{L_1 + \frac{M^4}{L_1 L_2 C_2 R_L^2}} \qquad (4.42)$$

The efficiency relates the power consumed by the load to the power delivered to all the resistances in the circuit. The capacitor C_1 does not impact on the computation of this metric so that the efficiency in the PS compensation topology has an identical definition to that derived for the SS compensation networks. Its expression can be found in Eq. 4.17.

As in the previous topologies, the bifurcation phenomenon must be avoided to work with a stable system. The procedure to determine how stable PS systems should be designed also relies on evaluating the number of roots (frequencies) of the reactance of Z_{in}. For this particular topology, the condition is expressed as [29]:

$$q_1 > q_2 \qquad (4.43)$$

The next graphs in Fig. 4.8 illustrate how the main electrical magnitudes vary with the frequency. It is important to note that the input current has a limited region in which the values do not exceed that associated with the designed resonant frequency. The configuration of the analysis is defined in Table 4.3.

The bifurcation analysis shows a different performance than systems with series compensation on the primary side. All the electrical magnitudes are kept below their nominal values except the primary current, which exceeds this value with frequency variations greater than about 5%, and the primary capacitor current, which increases linearly with frequency increments. Therefore, it is highly recommended that these systems have precise control of the current supplied by the source, which prevents damage to the primary components. The maximum charging power is reached in a very small range of frequency deviations. Except for these values, the performance of the system is relatively stable with and without bifurcation, although the designs that make it possible to avoid this phenomenon have a softer variation of the magnitudes except in current absorbed from the grid, which increases in a more pronounced way. The performance of the system is greater at all times in the systems without branches.

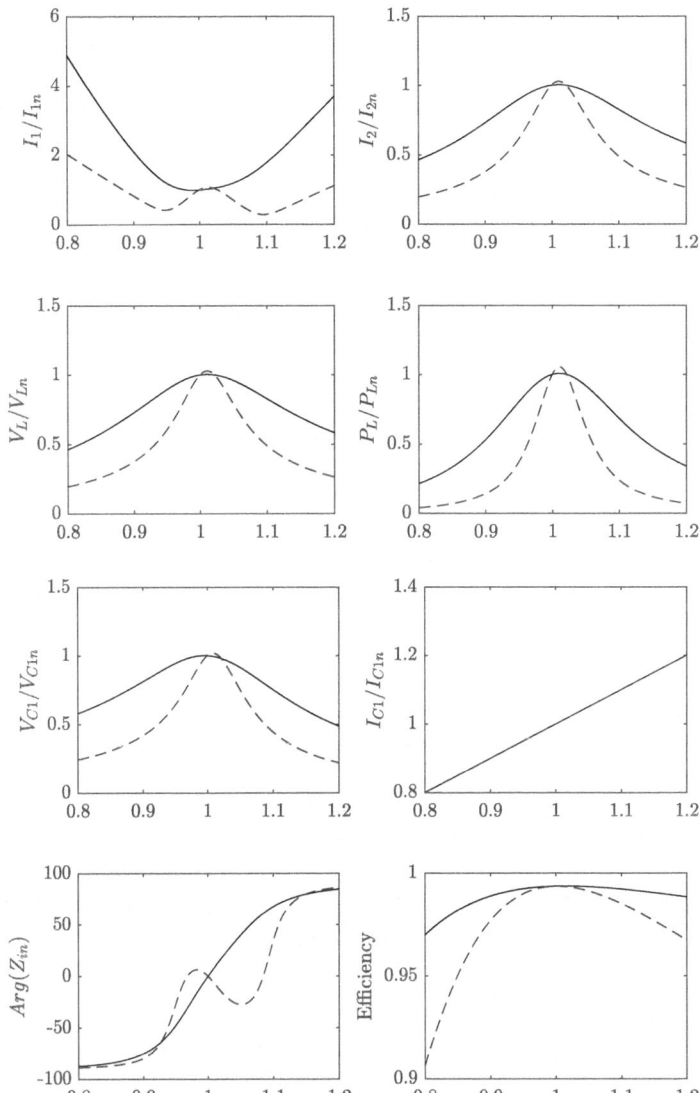

Fig. 4.8 Variation of the main electrical magnitudes for two magnetic resonance systems based on the PS compensation topology. Stable configuration is shown in black and the system with bifurcation is shown in blue

Table 4.3 Parameters used in the stability analysis with PS compensation network	Property	Non-bifurcation values (μH)	Bifurcation values (μH)
	L_1	12.4	4.3
	L_2	152	409
	M	6.3	6.1

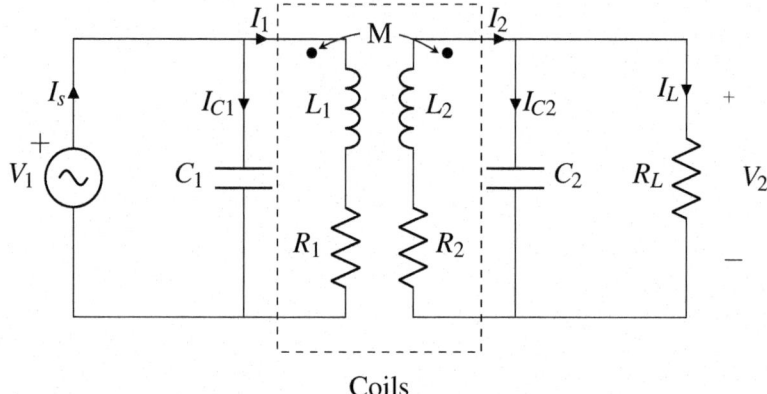

Coils

Fig. 4.9 Equivalent circuit of a magnetic resonance wireless charger with a Parallel–Parallel compensation topology

4.3.4 Parallel–Parallel Topology

The scheme of the Parallel–Parallel (PP) compensation topology is reflected in the Fig. 4.9.

The analysis of the primary and the secondary meshes gives the following expressions:

$$\mathbf{V_1} = (R_1 + j\omega L_1)\, \mathbf{I_1} - j\omega M \mathbf{I_2} \tag{4.44}$$

$$\mathbf{V_2} = -(R_2 + j\omega L_2+)\, \mathbf{I_2} + j\omega M \mathbf{I_1} = R_L \mathbf{I_L} \tag{4.45}$$

We know that $\mathbf{I_2} = \mathbf{I_{C2}} + \mathbf{I_L}$ and $\mathbf{I_s} = \mathbf{I_{C1}} + \mathbf{I_1}$. Because of the parallel connection of the capacitors, we can state that $\mathbf{I_{C1}} = j\omega C_1 \mathbf{V_1}$ and $\mathbf{I_{C2}} \frac{1}{j\omega C_2} = \mathbf{I_L} R_L$. After further processing, we derive that:

$$\mathbf{I_L} = \frac{j\omega M}{(R_2 + j\omega L_2)\,(1 + j\omega R_L C_2) + R_L}\mathbf{I_1} \tag{4.46}$$

$$\mathbf{I_2} = \frac{j\omega M\,(1 + j\omega R_L C_2)}{(R_2 + j\omega L_2)\,(1 + j\omega R_L C_2) + R_L}\mathbf{I_1} \tag{4.47}$$

These last expressions allow us to define the input impedance Z_{in} as follows:

$$Z_{in} = \frac{\mathbf{V_1}}{\mathbf{I_s}} = \frac{R_1 + j\omega L_1 + \frac{\omega^2 M^2 (1+j\omega R_L C_2)}{(R_2+j\omega L_2)(1+j\omega R_L C_2)+R_L}}{1 + j\omega C_1 \left(R_1 + j\omega L_1 + \frac{\omega^2 M^2 (1+j\omega R_L C_2)}{(R_2+j\omega L_2)(1+j\omega R_L C_2)+R_L}\right)} \tag{4.48}$$

where the imaginary part corresponds to:

$$Imag(Z_{in}) = \frac{\left(L_1\omega - \frac{\omega M^2}{L_2}\right)\left(1 - \omega C_1\left(L_1\omega - \frac{\omega M^2}{L_2}\right)\right) - \omega C_1\left(\frac{M^2 R_L}{L_2^2}\right)^2}{\left(1 - \omega C_1\left(L_1\omega - \frac{\omega M^2}{L_2}\right)\right)^2 - \omega^2 C_1^2\left(\frac{M^2 R_L}{L_2^2}\right)^2} \quad (4.49)$$

In order to transfer a high power to the load, the reactive component of the input impedance should be null. This condition implies that the capacitor C_1 should be set according to the following equation:

$$C_1 = \frac{C_2 L_2^2\left(L_1 L_2 - M^2\right)}{\left(L_1 L_2 - M^2\right)^2 + \frac{M^4 R_L^2 C_2}{L_2}} \quad (4.50)$$

As for the efficiency, this metric is approximated to:

$$\eta = \frac{R_L}{R_L + R_2 + \frac{R_1 L_2}{M^2} + \frac{R_2 R_L^2}{\omega_0^2 L_2} + \frac{R_1 R_L^2}{\omega_0^2 M^2}} \quad (4.51)$$

for the range of frequencies of usual magnetic resonance applications. Then, the efficiency is maximized when its denominator is minimized. This condition holds when:

$$\frac{R_2 R_L^2 M^2 + R_1 R_2 L_2^2}{\omega_0^2 M^2 L_2^2} \ll 1 \quad (4.52)$$

Under these circumstances, the optimum efficiency η_0 is expressed as:

$$\eta_0 = \frac{R_L}{\frac{R_1 L_2^2}{M^2} + R_2 + R_L} \quad (4.53)$$

From this last equation, we observe that it is convenient to design a PP wireless charger with a high mutual inductance but a low secondary self-inductance. For these systems, the bifurcation phenomenon is avoided when [29]:

$$q_1 > q_2 + \frac{1}{q_2} \quad (4.54)$$

The effects of the bifurcation phenomenon in a PP magnetic resonance system are illustrated in the set of graphs in Fig. 4.10. This analysis shows a similar system performance to that of the PS topology, although the difference between systems with and without bifurcation is less in this case. The configuration of the analysis is defined in Table 4.4.

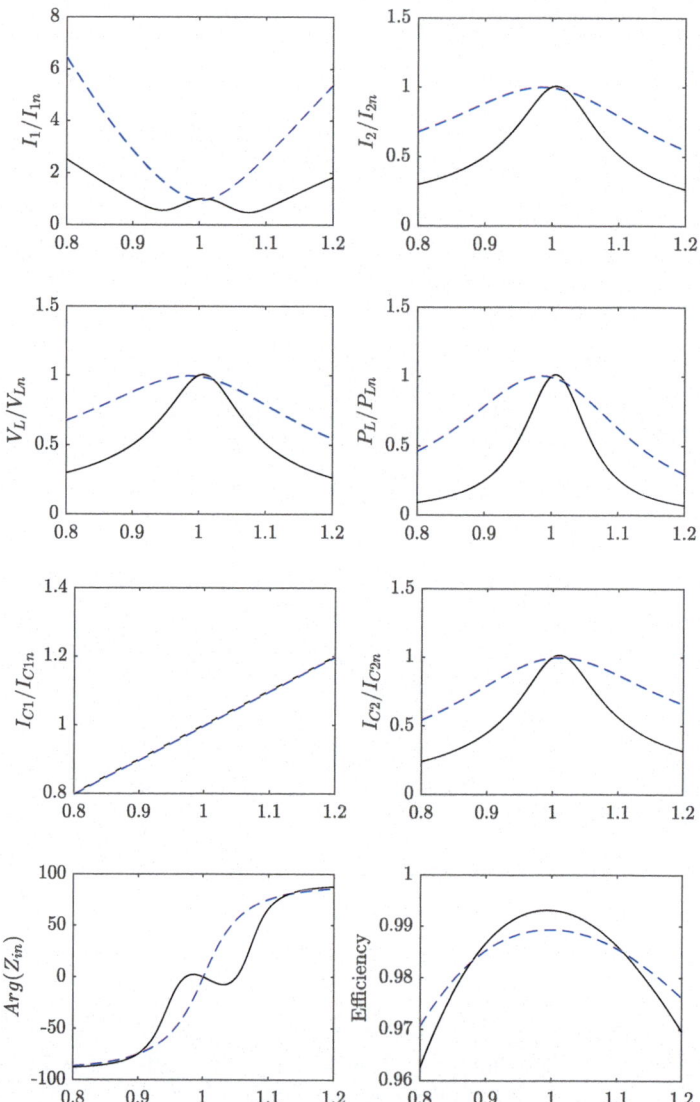

Fig. 4.10 Variation of the main electrical magnitudes for two magnetic resonance systems based on the PP compensation topology. Stable configuration is shown in black and the system with bifurcation is shown in blue

Table 4.4 Parameters used in the stability analysis with PP compensation network

Property	Non-bifurcation values (μH)	Bifurcation values (μH)
L_1	13.2	4.6
L_2	4.2	11.4
M	1.0	1.0

Table 4.5 Capacitor values, stability conditions and functionality of the four mono-resonant network topologies

	SS	SP	PS	PP
C_1	$\frac{1}{L_1\omega}$	$\frac{L_2^2 C_2}{L_1 L_2 - M^2}$	$\frac{L_2 C_2}{L_1 + \frac{M^4}{L_1 L_2 C_2 R_L^2}}$	$\frac{(L_1 L_2 - M^2) L_2^2 C_2}{\frac{M^4 R_L^2 C_2}{L_2^2}(L_1 L_2 - M^2)^2}$
C_2	$\frac{1}{L_2\omega}$	$\frac{1}{L_2\omega}$	$\frac{1}{L_2\omega}$	$\frac{1}{L_2\omega}$
Stability	$q_1 > \frac{4q_2^3}{4q_2^2 - 1}$	$q_1 > q_2 + \frac{1}{q_2}$	$q_1 > q_2$	$q_1 > q_2 + \frac{1}{q_2}$
Function	Voltage source	Voltage source	Current source	Current source

4.3.5 Discussion

Observing the design guidelines presented earlier, the SS topology is the only one in which the proper value of the primary capacitor is independent of the load impedance. The dependence of C_1 with R_L in the other three topologies represents an obstacle for practical applications.

For the mono-resonant compensation networks, it is the connection of the secondary capacitor which impacts on the system's behaviour. When this connection is in series, the charger acts as a voltage source delivering a constant voltage independently of the load value. On the other hand, when the secondary capacitor is connected in parallel with its corresponding coil, the system can be modelled as a current source.

As for the primary capacitor, its proper configuration depends on the connection of the secondary side. When the secondary capacitor is connected in series with the primary coil, the value of the primary capacitor should be related to the maximum voltage on the secondary coil. Otherwise, the value of the primary side is a function of the maximum current provided by the source.

Concerning the bifurcation phenomenon, the systems with a series compensation on the primary side are more stable than those with a primary compensation network in parallel. Moreover, the primary compensation networks in series prevent the electrical components from increasing the values of the voltage and/or current that they support. This provides an ideal protection for them. This condition does not hold with a primary compensation network in parallel. Specifically, for this kind of configuration the current flowing through the primary capacitor may exceed the current estimated in its designed operational frequency, i.e. the one forcing the system to work on resonance.

Another important difference in the behaviour of the compensation topologies lies in the grid current. Primary compensation networks in series maintain this metric without peaks. However, the connection in parallel leads to current peaks when the operation frequency greatly differs from the resonant one.

Table 4.5 summarises the main features of the four mono-resonant compensation topologies.

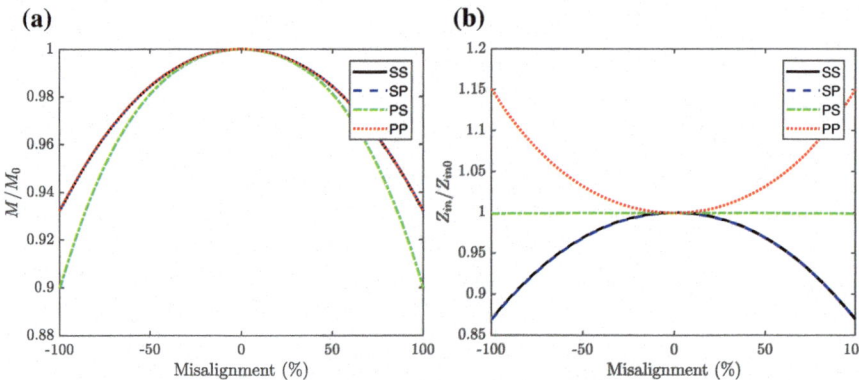

Fig. 4.11 Variation of M (**a**) and of $|Z_{in}|$ (**b**) with horizontal one-axis misalignment for the four mono-resonant compensation topologies

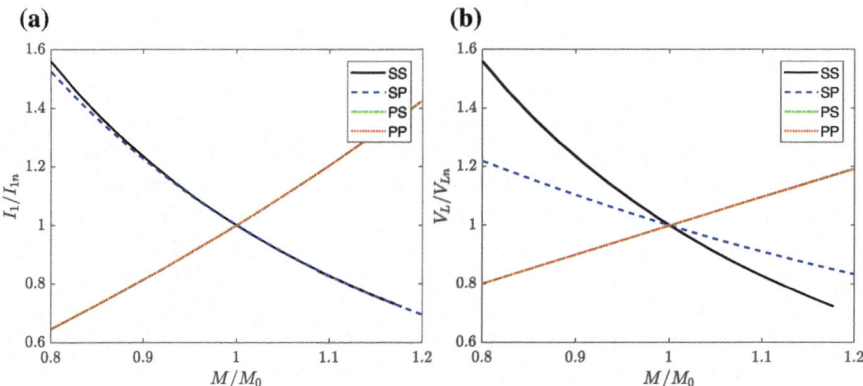

Fig. 4.12 Variation of I_1 (**a**) and of V_L (**b**) with misalignment for the four mono-resonant compensation topologies

In addition to bifurcation, tolerance to misalignment is key when deciding which topology should be implemented. Figure 4.11 shows how the mutual inductance and the equivalent impedance vary for different percentages of horizontal misalignment, assuming square coils are used. Please note that the coil self-inductances are different for every compensation topology as explained in the previous sections. It can be observed that the SS, SP and PP topologies have similar variations in their mutual inductance whereas the change in the mutual inductance is more abrupt in the PS topology. However, the PS topology has the highest tolerance of its equivalent input impedance to the degree of misalignment. In the SS and SP topologies, the input impedance decreases with coil misalignment whereas the PP topology suffers from increased input impedance under these circumstances.

Horizontal misalignment and a larger gap between the coils decreases the mutual inductance, whereas bringing the coils closer increases this parameter. Variations in the mutual inductance due to coil misalignment also impact on the values of other electrical magnitudes of the system. Of these, the primary current I_1 and the load voltage V_L are among the most relevant. The primary current affects the safe operation of the charger: an excessive primary current could exceed the current supported by the power devices. In a Series–Series compensation network, the current increases when the mutual inductance decreases. Figure 4.12 depicts these magnitudes for the four mono-resonant compensation topologies. The SS and PS topologies must include some control algorithms to prevent excessive voltage and currents [8, 20]. Alternatively, in the SP and PP topologies, a decrease in the mutual inductance leads to a decreased load voltage, which may not be sufficient to charge the battery correctly.

4.4 Multi-resonant Compensation Networks

The study of mono-resonant compensation networks has shown that they are capable of fulfilling the purpose for which they were designed under ideal conditions. However, wireless chargers almost never operate in these ideal conditions, due to certain factors (e.g. misalignments between coils, components whose characteristics are not ideal or variations in the functional environment) that compromise the system's reliability.

Given these factors, mono-resonant structures have not performed adequately. This behaviour has promoted the development of multi-resonant compensation structures, in which the compensation topologies are composed of more than one reactive component. As the analysis of some multi-resonant structures will demonstrate, they are capable of eliminating electrical problems such as bifurcation. In fact, the work in [28] shows how the Higher Order Compensation increases stability against disturbances and reduces the risk of producing the bifurcation phenomenon. They can also improve efficiency in bi-directional loaders even when coil misalignment occurs.

Multi-resonant structures can be designed in different ways. They may be proposed as basic mono-resonant configurations with the addition of extra capacitors. They can also be based on more complex topologies using coils in conjunction with new capacitors, relay resonator structures or multi-transmitter coils.

For electric vehicle charging applications, the number of suitable multi-resonant compensation structures is more limited due to the power involved in the power transfer. Some authors have limited the development of these structures to the inclusion of a greater number of capacitors to mono-resonant structures. Among these we can cite the SPS topology proposed in [27], in which the authors sought an intermediate behaviour between the SS and PS compensation topologies. Their goal was to achieve greater stability and a better behaviour in situations of misalignment than that provided by the mono-resonant compensation structures.

However, current researchers are focused on topologies composed of both coils and capacitors for the compensation networks. The use of both these components results in a wide variety of configurations depending on the number of components

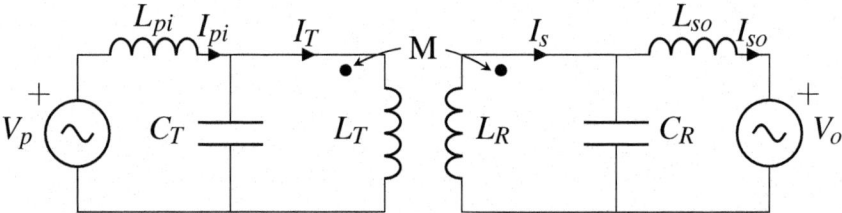

Fig. 4.13 LCL topology

used and their arrangement. In addition to the aforementioned improvements in stability and behaviour in terms of sensitivity to misalignment, the use of this type of compensation system provides a series of advantages over those composed solely of capacitors. These advantages include improved behaviour and efficiency in dynamic chargers [9, 31], feasibility of bi-directional systems [16] and capability of charging multiple vehicles simultaneously with a single primary coil [11]. The most relevant multi-resonant compensation networks are the inductor-capacitor-inductor (LCL) and the inductor-capacitor-capacitor (LCC) topologies, which are analysed next. As these are geared towards bi-directional chargers, the battery is modelled as an AC source in the equivalent circuit of the magnetic resonance charger. Please refer to Chap. 2 for an in-depth description.

4.4.1 LCL Topology

The simplest compensation topology comprising capacitors and coils is the LCL topology. This topology is usually composed of a coil connected in series to the output of the inverter and a capacitor connected in parallel to the transmission coil. When referring to LCL, the compensation networks on the primary and the secondary side are equivalent. Thus, for the secondary side, a capacitor is used in parallel to the receiver coil and a coil is connected to the input of the rectifier. The equivalent circuit of the LCL network is presented in Fig. 4.13.

When analysing this topology, we assume that the primary converter produces a sinusoidal voltage $V_p \angle 0°$ at an angular frequency ω. The voltage induced from the primary to the secondary side is:

$$\mathbf{V_s} = j\omega M \mathbf{I_T} \tag{4.55}$$

whereas the voltage reflected from the secondary side to the primary side corresponds to:

$$\mathbf{V_r} = -j\omega M \mathbf{I_s} \tag{4.56}$$

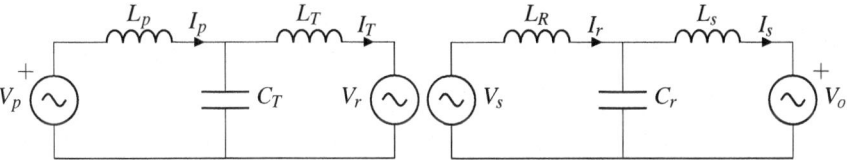

Fig. 4.14 Model for T-equivalent

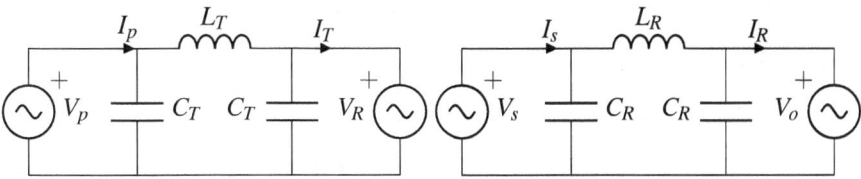

Fig. 4.15 Model for π-equivalent

Incorporating these voltages into the simplified circuit, we obtain an equivalent circuit of the LCL compensation topology as depicted in Fig. 4.14.

The LCL topology is built with the particularity that the coils used in the compensation network equal those used for the coupler so that $L_p = L_T$ and $L_R = L_S$. Thus, the equivalent circuit can be expressed in the π-model represented in Fig. 4.15.

If we apply Kirchhoff's Current Law on the primary side, we obtain:

$$\mathbf{I_p} = \mathbf{V_p} j\omega C_T + \frac{\mathbf{V_p} - \mathbf{V_r}}{j\omega L_T} \tag{4.57}$$

$$\mathbf{I_T} = \frac{\mathbf{V_p} - \mathbf{V_r}}{j\omega L_T} - j\omega C_T \mathbf{V_r} \tag{4.58}$$

For the resonant frequency ω_0, these two conditions hold:

$$\frac{1}{j\omega_0 L_T} - j\omega_0 C_T = 0 \tag{4.59}$$

$$\frac{1}{j\omega_0 L_R} - j\omega_0 C_R = 0 \tag{4.60}$$

So, Eqs. 4.57 and 4.58 are simplified as:

$$\mathbf{I_p} = j\frac{V_r}{\omega_0 L_T} \tag{4.61}$$

$$\mathbf{I_T} = -j\frac{V_p}{\omega_0 L_T} \tag{4.62}$$

The network on the secondary side is similar to the primary one, so with a circuital analysis is possible to derive that:

$$\mathbf{I_R} = j\frac{V_s}{\omega_0 L_R} \tag{4.63}$$

$$\mathbf{I_s} = -j\frac{V_o}{\omega_0 L_R} \tag{4.64}$$

The substitution of Eqs. 4.55 and 4.61 in Eq. 4.64 yields to:

$$\mathbf{I_s} = -j\frac{M \cdot V_p}{\omega_0 L_R L_T} \tag{4.65}$$

Analysing this equation, we observe that the system with a LCL compensation network behaves as a constant current source for a wide range of load variations. Consequently, the control required by the secondary side to charge the model is simple. This control will be based on the fact that the magnitude of this current source is related to the voltage obtained with the primary converter V_p. The work in [16] shows that this relationship holds even when there are multiple power receivers.

To control the power flow, we first need to derive the real power consumed by the secondary side P_s, which corresponds to:

$$P_s = \frac{M V_p}{\omega_0 L_R L_T} V_s sin(\theta) \tag{4.66}$$

with the phasor of the receiver voltage equalling $V_s \angle -\theta$. From this equation, it can be observed that the direction of the power flow can be controlled by θ. Positive values of θ imply a power flow from the grid to the battery. On the contrary, a negative θ leads to a power flow from the battery to the grid. The power level is adjusted by the magnitude V_s, obtaining the maximum peak when $\sin\theta$ is equal to 1.

The systems with the LCL topology are used in high power applications because they are able to offer a high transmission efficiency even with coils that have a low quality factor [28].

However, the LCL compensation network has two main disadvantages. Firstly, it only resonates at one frequency, which can limit the control approach to cope with coil misalignment. Secondly, the coils used in the compensation networks are the same as those used in the coupler. These are expensive coils, which will increase the costs of the charger.

4.4.2 LCC Topology

From the LCL topology, other multi-resonant topologies are derived. The LCC topology stands out among these on account of its popularity [14]. The diagram of this

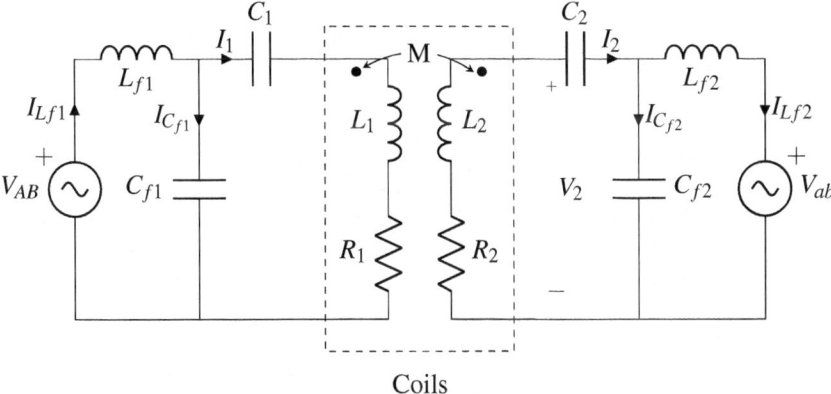

Fig. 4.16 Magnetic-resonant charger with LCC compensation

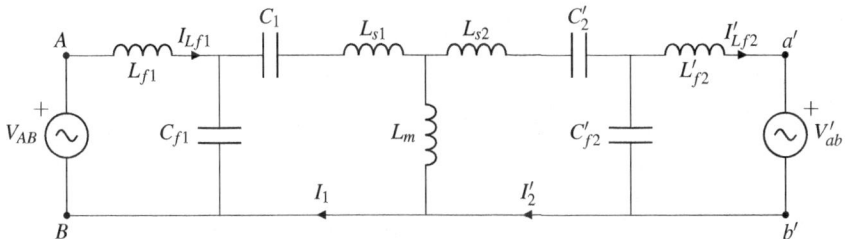

Fig. 4.17 T-model of the magneto-resonant charger with LCC compensation

topology is presented in Fig. 4.16. The output of the inverter is connected to an inductor. Two capacitors are also included in the compensation network.

LCC topology is normally used as the compensation network for both the primary and the secondary side. This configuration is referred to as the double-sided LCC or LCC-LCC [14]. However, there are other configurations for the secondary side as presented in [22]. With other types of compensation networks on the secondary side, the resonant frequency depends on the load features, so LCC-LCC is preferred in order to ease the control procedure.

As a result, we will analyse the LCC-LCC compensation topology in this chapter. To facilitate the electrical analysis of this configuration, the equivalent T-model of the coupled coils is derived first, as presented in Fig. 4.17.

By defining the variable n as:

$$n = \sqrt{\left(\frac{L_2}{L_1}\right)} \tag{4.67}$$

we can establish that:

$$L_m = k \cdot L_1 \tag{4.68}$$

$$L_{s1} = (1 - k) L_1 \tag{4.69}$$

$$L'_{s2} = (1 - k) \frac{L_2}{n^2} \tag{4.70}$$

$$L'_{f2} = \frac{L_{f2}}{n^2} \tag{4.71}$$

$$C'_2 = n^2 \cdot C_2 \tag{4.72}$$

$$C'_{f2} = n^2 \cdot C_{f2} \tag{4.73}$$

$$\mathbf{V}'_{ab} = \frac{\mathbf{V}_{ab}}{n} \tag{4.74}$$

To make the system work on resonance at frequency ω_0, we impose the following four conditions:

$$L_{f1}C_{f1} = \frac{1}{\omega_0^2} \tag{4.75}$$

$$L_{f2}C_{f2} = \frac{1}{\omega_0^2} \tag{4.76}$$

$$\left(L_1 - L_{f1}\right) C_1 = \frac{1}{\omega_0^2} \tag{4.77}$$

$$\left(L_2 - L_{f2}\right) C_2 = \frac{1}{\omega_0^2} \tag{4.78}$$

The components in series are simplified with an equivalent inductance. In particular, the capacitor C_1 and the inductance L_{s1} are modelled with the inductance L_{e1}. Similarly, the capacitor C' and the inductance L'_{s2} are reduced to L'_{e2}. The expressions for these two new inductors are:

$$L_{e1} = \frac{1}{j\omega_0} \left(\frac{1}{j\omega_0 C_1} + j\omega_0 L_{s1} \right) = L_{f1} - k \cdot L_1 \tag{4.79}$$

$$L'_{e2} = \frac{1}{j\omega_0} \left(\frac{1}{j\omega_0 C'_2} + j\omega_0 L'_{s2} \right) = L'_{f2} - k \cdot L_1 \tag{4.80}$$

The superposition theorem is then applied for the electrical analysis so that the effects of the sources $\mathbf{V}_{AB}$ and $\mathbf{V}'_{AB}$ are computed independently.

When only the source $\mathbf{V}_{AB}$ is considered, L'_{f2} and C'_{f2} are in parallel. At the resonant frequency, these two elements behave as an open circuit so that $\mathbf{I}'_{2AB} = 0$. For the other components, we can establish that:

$$L_{e1} + L_m = L_{f1} - k \cdot L_1 + k \cdot L_1 = L_{f1} \tag{4.81}$$

The capacitor C_{f1} is in parallel with the series association of L_{e1} and L_m. Considering the condition imposed by Eq. 4.75, at the resonant frequency, these elements are also an open circuit so that $\mathbf{I}_{Lf1AB} = 0$.

From this previous analysis, we can conclude that the voltage in C_{f1} is $\mathbf{V}_{AB}$ and the voltage in C'_{f2} is equivalent to the voltage in L_m. The currents $\mathbf{I}_{AB}$ and $\mathbf{I}'_{Lf2AB}$ can be expressed as:

$$\mathbf{I}_{1AB} = \frac{\mathbf{V}_{AB}}{j\omega_0 L_{f1}} \tag{4.82}$$

$$\mathbf{I}'_{Lf2AB} = \frac{\mathbf{V}_{AB} - j\omega_0 L_{e1} I_{1AB}}{j\omega_0 L_{f1}} = \frac{k \cdot \mathbf{V}_{AB} \cdot L_1}{j\omega_0 L_{f1} L'_{f2}} \tag{4.83}$$

The analysis when only the source $\mathbf{V}'_{ab}$ is active is analogous. This yields to:

$$\mathbf{I}_{1ab} = 0 \tag{4.84}$$

$$\mathbf{I}'_{2ab} = -\frac{\mathbf{V}'_{ab}}{j\omega_0 L'_{f2}} \tag{4.85}$$

$$\mathbf{I}_{Lf1ab} = -\frac{k \cdot \mathbf{V}'_{ab} \cdot L_1}{j\omega_0 L_{f1} L'_{f2}} \tag{4.86}$$

$$\mathbf{I}'_{Lf2ab} = 0 \tag{4.87}$$

Since $\mathbf{V}'_{ab}$ is the input voltage at the secondary power converter, it will have the same phase as $\mathbf{I}'_{Lf2}$. As $\mathbf{I}'_{Lf2ab} = 0$, $\mathbf{V}'_{ab}$ is in phase with $\mathbf{I}'_{Lf2AB}$. Considering $\mathbf{U}_{AB}$ as the reference for the phasor angles, we obtain that:

$$\mathbf{V}_{AB} = V_{AB}\underline{/0°} \tag{4.88}$$

$$\mathbf{V}'_{ab} = V'_{ab}\underline{/-90o} \tag{4.89}$$

and the currents are:

$$\mathbf{I_{Lf1}} = \mathbf{I_{Lf1ab}} = \frac{k \cdot V'_{ab} \cdot L_1}{\omega_0 L_{f1} L'_{f2}} \underline{/0°} = \frac{k\sqrt{L_1 L_2} V_{ab}}{\omega_0 L_{f1} L_{f2}} \underline{/0°} \tag{4.90}$$

$$\mathbf{I_{Lf1}} = \mathbf{I_{Lf1ab}} = \frac{k \cdot V'_{ab} \cdot L_1}{\omega_0 L_{f1} L'_{f2}} \underline{/0°} = \frac{k\sqrt{L_1 L_2} V_{ab}}{\omega_0 L_{f1} L_{f2}} \underline{/0°} \tag{4.91}$$

$$\mathbf{I_1} = \mathbf{I_{1AB}} = \frac{V_{AB}}{j\omega_0 L_{f1}} = \frac{V_{AB}}{\omega_0 L_{f1}} \underline{/90°} \tag{4.92}$$

$$\mathbf{I_2} = \frac{\mathbf{I'_2}}{n} = \frac{\mathbf{I'_{2ab}}}{n} = \frac{V_{ab}}{\omega_0 L_{f2}} \underline{/0°} \tag{4.93}$$

$$\mathbf{I_{Lf2}} = \mathbf{I'_{Lf2}} = \frac{\mathbf{I'_{Lf2AB}}}{n} = \frac{k\sqrt{L_1 L_2} \mathbf{V_{AB}}}{\omega_0 L_{f1} L_{f2}} \underline{/90°} \tag{4.94}$$

As the power factor is the unity, the power transferred to the battery is computed as:

$$P = \mathbf{V_{AB}} \mathbf{I_{Lf1}} = k \frac{\sqrt{L_1 L_2}}{\omega_0 L_{f1} L_{f2}} V_{AB} V_{ab} \cos\phi \tag{4.95}$$

To configure an LCC-LCC compensation topology, we set that L_{f1} and L_{f2} are equal. In particular, they will be designed according to the following expression:

$$L_{f1} = L_{f2} = \mathbf{V_{AB}} \mathbf{I_{Lf1}} = \sqrt{k \frac{\sqrt{L_1 L_2}}{\omega_0 P} V_{AB} V_{ab} \cos\phi} \tag{4.96}$$

The values for the capacitors C_{f1} and C_{f2} are computed based on Eqs. 4.75 and 4.76. Alternatively, the values for C_1 and C_2 are derived from Eqs. 4.77 and 4.78 as follows:

$$C_{f1} = \frac{1}{\omega_0^2 L_{f1}} \tag{4.97}$$

$$C_{f2} = \frac{1}{\omega_0^2 L_{f2}} \tag{4.98}$$

$$C_1 = \frac{1}{\omega_0^2 (L_1 - L_{f1})} \tag{4.99}$$

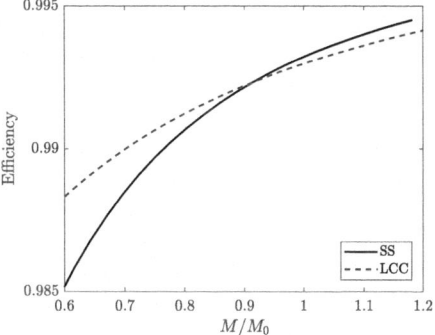

Fig. 4.18 Comparison of efficiency between SS and LCL topology

$$C_2 = \frac{1}{\omega_0^2 \left(L_2 - L_{f2}\right)} \tag{4.100}$$

It is of interest to compare this multi-resonant topology (as it is one of the most widely used) with the most common mono-resonant topology, the SS. The fact of using a greater number of components in the compensation system affects the efficiency of the system. Systems with SS compensation, using only a series capacitor connected on both sides of the charger, have lower losses than LCC systems, which use three components on both sides. The maximum efficiency of the system is, therefore, greater in the SS compensation topology than in the LCC.

However, the real difference between the two topologies does not lie in the maximum efficiency that the system achieves with them, but in the behaviour that each topology provides before factors such as misalignment and the variation of the equivalent impedance of the battery [5, 7, 23, 30] are taken into account. Figure 4.18 shows a comparison of the efficiency of a wireless charger with both topologies. It shows how the maximum efficiency of the SS system is concentrated in the centre of the curve, where there is no misalignment, with a value higher than that achieved by the LCC topology. Although the LCC topology has a lower maximum efficiency, it remains practically constant with a greater range of misalignments in both axes.

In the same way, this difference is also noticeable in the evolution of the power factor when faced with misalignments, although in this case the effect is greater than with efficiency. With both topologies, the unit power factor is reached but the LCC topology is capable of maintaining high values in a wide range of misalignments.

The selected topology also affects the currents and voltages of the system components [13]. Figure 4.19 shows the difference before variations of the mutual inductance. In (a), we show how the voltage of the capacitor with a series-connection in the LCC topology is inferior to that of the SS topology anyway, and, in addition, this value is less influenced by mutual inductance variations, which reduce the requirements of the capacitors. In the same way, the currents that circulate through the primary (b) are lower with the LCC topology, requiring conductors with smaller cross-sections and, therefore, of lower cost.

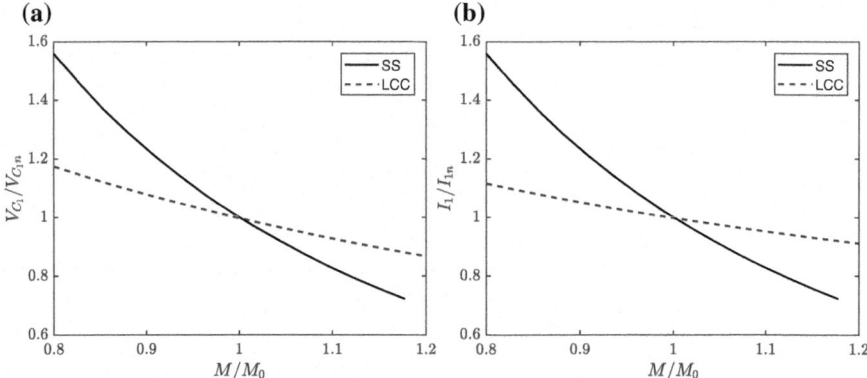

Fig. 4.19 Comparison of **a** the voltage of the capacitor connected in series and of **b** the currents that circulate through the primary between SS and LCL topology

The LCC topology also provides the secondary side of the charger with a very practical functionality by reducing the instabilities created by the rectifier [11]. These instabilities are due to the non-linear behaviour of the rectifiers, which is more pronounced when additional power electronics are used behind the rectifier to control the charge. The additional coil provided by this topology reduces these effects.

The LCC topology is also recommended for dynamic charging systems [6, 31], with multiple secondary coils [11, 30], and for bi-directional systems with grid injection [16, 18, 19].

Both for dynamic loading operations and with multiple receiving coils, larger primary coils are used, usually known as track coils. In order to comply with the characteristics of these systems, it is usual to design the system to circulate a constant current through the primary coil [2, 3]. This constant current condition is satisfied in the LCC topology when its design is correctly performed so that it resonates to the switching frequency [30].

Although the LCC topology is the most used for charging applications, there are also other less used that provide specific improvements. These include the topology LCL-S [1], which is a simpler topology because the secondary is only composed of a capacitor in series with the coil. The LCL-S topology allows the charger to behave like a voltage source before the battery. The topology Capacitor-Inductor-Capacitor-Inductor (CLCL) [25] is used in bi-directional systems.

References

1. Cai, C., Wang, J., Fang, Z., Zhang, P., Hu, M., Zhang, J., Li, L., Lin, Z.: Design and optimization of load-independent magnetic resonant wireless charging system for electric vehicles. IEEE Access **6**, 17264–17274 (2018). https://doi.org/10.1109/ACCESS.2018.2810128, http://ieeexplore.ieee.org/document/8303685/

2. Covic, G.A., Boys, J.T.: Inductive power transfer. Proc. IEEE **101**(6), 1276–1289 (2013). https://doi.org/10.1109/JPROC.2013.2244536, http://ieeexplore.ieee.org/document/6492113/

3. Covic, G.A., Boys, J.T.: Modern trends in inductive power transfer for transportation applications. IEEE J. Emerg. Sel. Top. Power Electron. **1**(1), 28–41 (2013). https://doi.org/10.1109/JESTPE.2013.2264473, http://ieeexplore.ieee.org/document/6517868/

4. Di Capua, G., Sanchez, J.A.A., Cabrera, A.T., Cabrera, D.F., Femia, N., Petrone, G., Spagnuolo, G.: A losses-based analysis for electric vehicle wireless chargers. In: 2015 International Conference on Synthesis, Modeling, Analysis and Simulation Methods and Applications to Circuit Design (SMACD), pp. 1–4. IEEE (2015). https://doi.org/10.1109/SMACD.2015.7301677, http://ieeexplore.ieee.org/document/7301677/

5. Esteban, B., Sid-Ahmed, M., Kar, N.C.: A comparative study of power supply architectures in wireless EV charging systems. IEEE Trans. Power Electron. **30**(11), 6408–6422 (2015). https://doi.org/10.1109/TPEL.2015.2440256, http://ieeexplore.ieee.org/document/7119591/

6. Feng, H., Cai, T., Duan, S., Zhao, J., Zhang, X., Chen, C.: An LCC-compensated resonant converter optimized for robust reaction to large coupling variation in dynamic wireless power transfer. IEEE Trans. Ind. Electron. **63**(10), 6591–6601 (2016). https://doi.org/10.1109/TIE.2016.2589922, http://ieeexplore.ieee.org/document/7508947/

7. Ge, S., Liu, C., Li, H., Guo, Y., Cai, G.: Double-LCL resonant compensation network for electric vehicles wireless power transfer: experimental study and analysis. IET Power Electron. **9**(11), 2262–2270 (2016). https://doi.org/10.1049/iet-pel.2015.0186, https://digital-library.theiet.org/content/journals/10.1049/iet-pel.2015.0186

8. González-González, J.M., Triviño-Cabrera, A., Aguado, J.A., González-González, J.M., Triviño-Cabrera, A., Aguado, J.A.: Design and validation of a control algorithm for a SAE J2954-compliant wireless charger to guarantee the operational electrical constraints. Energies **11**(3), 604 (2018). https://doi.org/10.3390/en11030604, http://www.mdpi.com/1996-1073/11/3/604

9. Hao, H., Covic, G.A., Boys, J.T.: An approximate dynamic model of LCL-T-based inductive power transfer power supplies. IEEE Trans. Power Electron. **29**(10), 5554–5567 (2014). https://doi.org/10.1109/TPEL.2013.2293138, http://ieeexplore.ieee.org/document/6676797/

10. Ibrahim, M.: Wireless Inductive Charging for Electrical Vehicules: Electromagnetic Modelling and Interoperability Analysis. Ph.D. thesis (2014). https://tel.archives-ouvertes.fr/tel-01127163

11. Keeling, N., Covic, G., Boys, J.: A unity-power-factor IPT pickup for high-power applications. IEEE Trans. Ind. Electron. **57**(2), 744–751 (2010). https://doi.org/10.1109/TIE.2009.2027255, http://ieeexplore.ieee.org/document/5173523/

12. Lee, B.-S., Han, K.-H.: Modeling and analysis of IPT system used for PRT. In: 2005 International Conference on Electrical Machines and Systems, pp. 839–842. IEEE (2005). https://doi.org/10.1109/ICEMS.2005.202656, http://ieeexplore.ieee.org/document/1574889/

13. Li, W., Zhao, H., Deng, J., Li, S., Mi, C.C.: Comparison study on SS and double-sided LCC compensation topologies for EV/PHEV wireless chargers. IEEE Trans. Veh. Technol. **65**(6), 4429–4439 (2016). https://doi.org/10.1109/TVT.2015.2479938, http://ieeexplore.ieee.org/document/7277103/

14. Li, W., Zhao, H., Li, S., Deng, J., Kan, T., Mi, C.C.: Integrated LCC compensation topology for wireless charger in electric and plug-in electric vehicles. IEEE Trans. Ind. Electron. **62**(7), 4215–4225 (2015). https://doi.org/10.1109/TIE.2014.2384003, http://ieeexplore.ieee.org/document/6991579/

15. Madawala, U.K., Neath, M., Thrimawithana, D.J.: A power frequency controller for bidirectional inductive power transfer systems. IEEE Trans. Ind. Electron. **60**(1), 310–317 (2013). https://doi.org/10.1109/TIE.2011.2174537, http://ieeexplore.ieee.org/document/6068247/

16. Madawala, U.K., Thrimawithana, D.J.: A bidirectional inductive power interface for electric vehicles in V2G systems. IEEE Trans. Ind. Electron. **58**(10), 4789–4796 (2011). https://doi.org/10.1109/TIE.2011.2114312, http://ieeexplore.ieee.org/document/5711663/

17. Moghaddami, M., Sarwat, A.: Self-tuning variable frequency controller for inductive electric vehicle charging with multiple power levels. IEEE Trans. Transp. Electrific. **3**(2), 488–

495 (2017). https://doi.org/10.1109/TTE.2016.2638642, http://ieeexplore.ieee.org/document/7781643/

18. Mohamed, A.A.S., Berzoy, A., de Almeida, F.G.N., Mohammed, O.: Modeling and assessment analysis of various compensation topologies in bidirectional IWPT system for EV applications. IEEE Trans. Ind. Appl. **53**(5), 4973–4984 (2017). https://doi.org/10.1109/TIA.2017.2700281, http://ieeexplore.ieee.org/document/7917270/

19. Mohamed, A.A.S., Marim, A.A., Mohammed, O.A.: Magnetic design considerations of bidirectional inductive wireless power transfer system for EV applications. IEEE Trans. Magn. **53**(6), 1–5 (2017). https://doi.org/10.1109/TMAG.2017.2656819, http://ieeexplore.ieee.org/document/7829389/

20. Murliky, L., Porto, R.W., Brusamarello, V.J., Sousa, F.R.d., Trivino-Cabrera, A., Azambuja, R.: Robust active tuning for wireless power transfer to support misalignments and variable load. In: 2018 IEEE Wireless Power Transfer Conference (WPTC), pp. 1–4. IEEE (2018). https://doi.org/10.1109/WPT.2018.8639439, https://ieeexplore.ieee.org/document/8639439/

21. Niu, W.Q., Chu, J.X., Gu, W., Shen, A.D.: Exact analysis of frequency splitting phenomena of contactless power transfer systems. IEEE Trans. Circuits Syst. I: Reg. Pap. **60**(6), 1670–1677 (2013). https://doi.org/10.1109/TCSI.2012.2221172, http://ieeexplore.ieee.org/document/6363491/

22. Pantic, Z., Bai, S., Lukic, S.M.: ZCS LCC-compensated resonant inverter for inductive-power-transfer application. IEEE Trans. Ind. Electron. **58**(8), 3500–3510 (2011). https://doi.org/10.1109/TIE.2010.2081954, http://ieeexplore.ieee.org/document/5587891/

23. Park, M., Nguyen, V.T., Yu, S.-D., Yim, S.-W., Park, K., Min, B.D., Kim, S.-D., Cho, J.G.: A study of wireless power transfer topologies for 3.3 kW and 6.6 kW electric vehicle charging infrastructure. In: 2016 IEEE Transportation Electrification Conference and Expo, Asia-Pacific (ITEC Asia-Pacific), pp. 689–692. IEEE (2016). https://doi.org/10.1109/ITEC-AP.2016.7513041, http://ieeexplore.ieee.org/document/7513041/

24. Sallan, J., Villa, J., Llombart, A., Sanz, J.: Optimal design of ICPT systems applied to electric vehicle battery charge. IEEE Trans. Ind. Electron. **56**(6), 2140–2149 (2009). https://doi.org/10.1109/TIE.2009.2015359, http://ieeexplore.ieee.org/document/4787071/

25. Thrimawithana, D.J., Madawala, U.K.: A generalized steady-state model for bidirectional IPT systems. IEEE Trans. Power Electron. **28**(10), 4681–4689 (2013). https://doi.org/10.1109/TPEL.2012.2237416, http://ieeexplore.ieee.org/document/6401199/

26. Triviño-Cabrera, A., Aguado, J.A.: A review on the fundamentals and practical implementation details of strongly coupled magnetic resonant technology for wireless power transfer. Energies **11**(10), 2844 (2018). https://doi.org/10.3390/en11102844, http://www.mdpi.com/1996-1073/11/10/2844

27. Villa, J.L., Sallan, J., Sanz Osorio, J.F., Llombart, A.: High-misalignment tolerant compensation topology for ICPT systems. IEEE Trans. Ind. Electron. **59**(2), 945–951 (2012). https://doi.org/10.1109/TIE.2011.2161055, http://ieeexplore.ieee.org/document/5985526/

28. Wang, C.S., Covic, G., Stielau, O.: Investigating an LCL load resonant inverter for inductive power transfer applications. IEEE Trans. Power Electron. **19**(4), 995–1002 (2004). https://doi.org/10.1109/TPEL.2004.830098, http://ieeexplore.ieee.org/document/1310386/

29. Wang, C.S., Covic, G., Stielau, O.: Power transfer capability and bifurcation phenomena of loosely coupled inductive power transfer systems. IEEE Trans. Ind. Electron. **51**(1), 148–157 (2004). https://doi.org/10.1109/TIE.2003.822038, http://ieeexplore.ieee.org/document/1265794/

30. Zhang, W., Mi, C.C.: Compensation topologies of high-power wireless power transfer systems. IEEE Trans. Veh. Technol. **65**(6), 4768–4778 (2016). https://doi.org/10.1109/TVT.2015.2454292, http://ieeexplore.ieee.org/document/7152988/

31. Zhu, Q., Wang, L., Guo, Y., Liao, C., Li, F.: Applying LCC compensation network to dynamic wireless EV charging system. IEEE Trans. Ind. Electron. **63**(10), 6557–6567 (2016). https://doi.org/10.1109/TIE.2016.2529561, http://ieeexplore.ieee.org/document/7405298/

Chapter 5
Power Electronics

5.1 Power Electronics

A charger needs to adapt the power provided by the external source to the electrical requirements of the EV battery. Moreover, in order to improve the power transfer through the coupled coils, an increase in the operational frequency is required for WPT implementations. These conversions are carried out efficiently by power converters so that they can adjust the voltage levels, their frequency, the currents and/or the impedances with reduced losses. Power converters are composed of semiconductors configured to operate as switches varying their ideal state (open or closed) at a switching frequency. Figure 5.1 reflects the traditional classification of power converters according to the power flow and frequency of the input and the output signal.

Specifically, several power converters are included in WPT chargers. Figure 5.2 reflects these structures in a magnetic resonance EV charger. The upper row converts the frequency of the signal provided by the grid from 50/60 Hz to a high frequency. For this operation, two steps are required. The first one consists in rectifying the grid power to obtain a DC voltage stored in a DC link. The DC link is then connected to DC/AC converter, which is able to generate an AC voltage at a high frequency. In contrast to the voltage provided by the grid, the signal obtained from the primary DC/AC converter may not be sinusoidal. The DC/AC output excites the coupled coils, so that an AC voltage is induced on the secondary side. As the EV battery must be fed with a DC signal, power converters adapt the voltage accordingly. An AC/DC converter is the first component connected to the secondary compensation network. If necessary, the voltage level is adjusted to the battery requirement with a DC/DC converter.

Power converters should be bi-directional in V2G applications to allow the power flow from the grid to the battery and vice versa.

The type and configuration of the power converters are mainly decided according to their functionality and the level of managed power, but other parameters such as weight, costs and electromagnetic emissions also affect the decision. For all of them, their efficiency is used as a metric to characterise their performance. Efficiency is defined in Eq. 5.1 as the quotient of the output real power P_{out} divided by the input

© Springer Nature Switzerland AG 2020

A. Triviño-Cabrera et al., *Wireless Power Transfer for Electric Vehicles: Foundations and Design Approach*, Power Systems,
https://doi.org/10.1007/978-3-030-26706-3_5

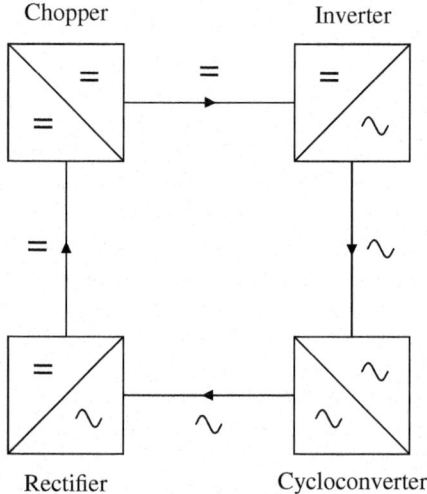

Fig. 5.1 Classification of power converters

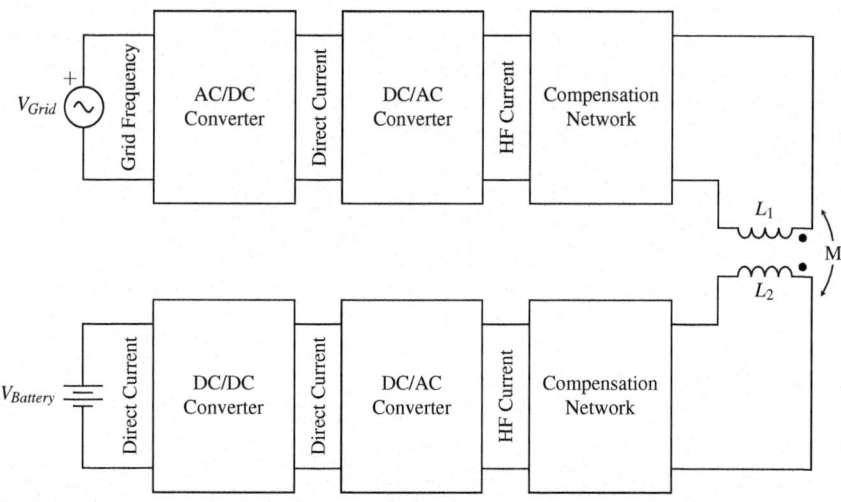

Fig. 5.2 Generic structure of an EV magnetic resonance wireless charger

real power P_{in}. The difference of these two values corresponds to the losses P_{losses}. These losses originate in the fact that the semiconductors are not ideal switches but present some internal losses when acting as an open circuit or a short-circuit and during the transitions of these states. Some power converters also rely on reactive components, which increase the losses.

$$\eta = \frac{P_{out}}{P_{in}} = \frac{P_{out}}{P_{out} + P_{losses}} \quad (5.1)$$

Ideally, a power converter should operate with an efficiency closed to unity. To achieve this goal, the components of the power converters should be carefully selected according to the voltage and current they should support and their switching frequency. The configuration of these components can be diverse, which has led to the proposal of multiple topologies for power converters. The suitability of a topology is related to the requirements of the power conversion and its control capabilities.

This chapter reviews the main issues regarding the power converters of wireless EV chargers. Firstly, a study of the power converter components is presented. Then, the configuration and control of power converters are described for uni-directional and bi-directional chargers.

5.2 Semiconductor Devices for Power Converters

Power electronics are used for multiple types of applications such as renewable energy converters, electronics or wireless chargers. The basis for power converters relies on using semiconductors as switching devices.

Currently there are several options for semiconductors that can be used in power converters. These include the Insulated Gate Bipolar Transistor (IGBT), the Metal Oxide Semiconductor Field Effect Transistor (MOSFET), the Gate Turn-off Thyristor (GTO), the diode and the thyristor. These components are depicted in Fig. 5.3.

Their suitability depends on the switching frequency and the voltage and/or current they should support. Figure 5.4 illustrates the regions of viable operation of the semiconductors.

Magnetic resonance chargers operate at high frequency, which implies that the semiconductor devices in power converters must switch at those frequencies (above 20 kHz). With this restriction, only IGBT and MOSFET are the valid options for power converters. IGBT are close to their frequency limit so their use in EV wireless chargers leads to noticeable losses due to their switching. They would also require refrigeration systems. Conventional Silicon MOSFET can switch at higher frequencies but the power they support is much more reduced.

Silicon Carbide (SiC) technology [25] overcomes this limitation with regard to MOSFET. Due to a wider bandgap, the SiC devices acquire four main superior material properties [21]:

Fig. 5.3 Representation of the most popular semiconductor devices in power converters

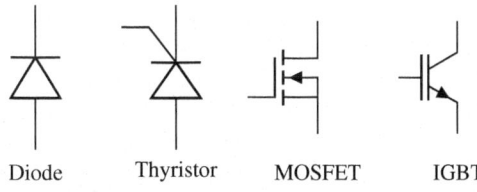

Diode   Thyristor   MOSFET   IGBT

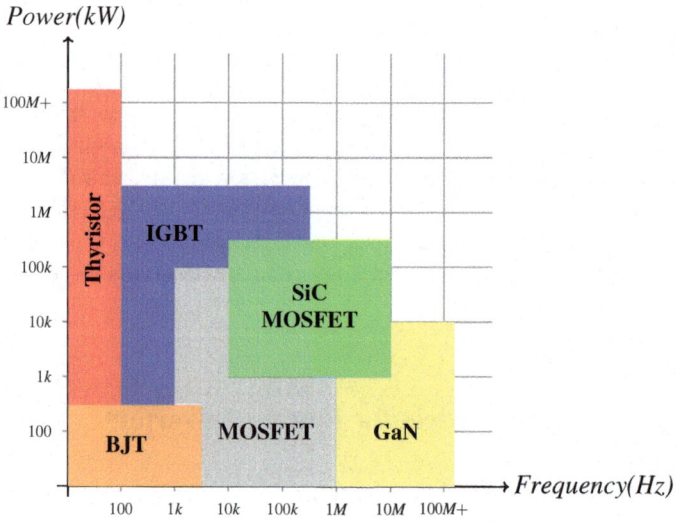

Fig. 5.4 Regions of viable operation for the controllable semiconductor devices

1. **Higher breakdown voltage**. It supports higher voltages.
2. **Lower leakage currents**. Since the thermal generation of electron-hole pairs is significantly slower than in silicon semiconductors, currents in off-state are also smaller and, in turn, losses are reduced as well.
3. **Higher thermal conductivity**. Heat can be dissipated in a more efficient way.
4. **Lower on-state resistance**, which reduces the losses when the device is active.

SiC technology has been applied to diodes, MOSFET, IGBT and Bipolar Junction Transistor (BJT). SiC MOSFETs are the most developed ones [33]. Despite its benefits, there is a limited number of manufacturers producing this kind of device. However, interest in this technology is expected to increase in the near future, which will boost its commercialisation [19].

Thus, when designing a wireless charger, the components of the power converters must be selected according to the operational frequency and the power they have to support. In addition, the snubber circuits must be designed, which are relevant systems that help the semiconductor devices to operate during the switchings. As explained previously, the semiconductor devices operate as switches in the power converters so that they commute from the open to the closed state and vice-versa at the switching frequency. These transitions are not instantaneous and some losses occur as shown in Fig. 5.5 for a generic MOSFET. In addition to the losses and related heat, these changes in the magnitudes di/dt and dv/dt infer stress in the components and excessive EMI. Snubber circuits lessen these effects. Snubber circuits are composed of active or passive elements that help to reduce electrical variables during the transient state. Figure 5.6 shows two representative configurations of snubber circuits.

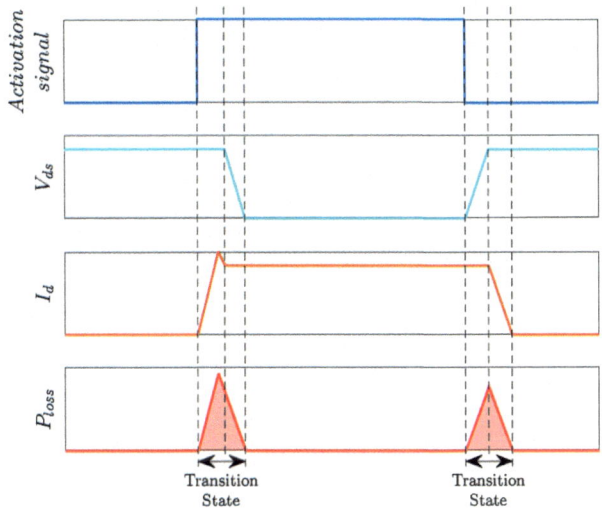

Fig. 5.5 Hard Switching transitions

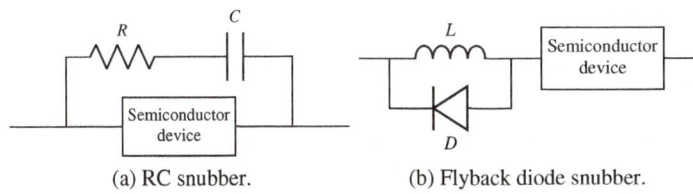

(a) RC snubber.   (b) Flyback diode snubber.

Fig. 5.6 Snubber basic topologies

In order to reduce losses, several techniques have been developed to configure the conditions in which the semiconductor devices should switch. Control systems are necessary for the power converters in order to configure when their semiconductor devices will switch. This decision may be supported by measurements taken from the converter output, the aim of the controller being to force the converter to generate an output signal with specific features (the reference levels). According to these measurements, the controller generates the switching signals for the power converters. Figure 5.7 depicts a generic diagram of the control system associated with a power converter.

The measurements used by the controller may come from the output of the power converter it is managing or from other electrical components of the system. For instance, in a magnetic resonance charger, the controller on the primary side usually depends on the status of the battery, so it uses measurements from the secondary side to configure the switching signals. The measurements are then sent through a wireless channel.

In complex systems, such as a magnetic resonance charger, there may be more than one controllable power converter. The power converter controllers may work

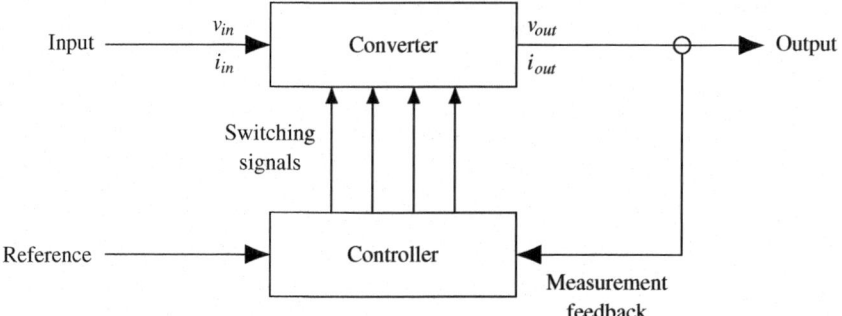

Fig. 5.7 Generic operation of a controller

individually or in a cooperative way exchanging information through a communication system.

5.3   Uni-directional Charger

Uni-directional chargers are designed to allow a power flow from the grid to the battery aimed at operating according to a charging mode (mainly CC, CV, CC-CV). Thus, the controller should receive these electrical magnitudes with a reduced delay in order to determine the switching times. The measurements may be derived from the BMS or from sensors specifically installed in the charger for this kind of operation.

The control of the magnetic resonance charger can be established by one of these three main approaches:

- **Primary-side controlled charger**. The complexity of the control system is placed on the primary side, because some data from the secondary side is needed to determine how to act. Under this approach, the on-board electronics and their corresponding weight are minimised. Figure 5.8 reflects this type of control.
- **Secondary-side controlled charger**. The controller acquiring the related-battery measurements is on the secondary side. The communication system is avoided as the controller is connected to the sensors performing the measurements. However, the complexity of the secondary side is greater, with more electronics and control, which leads to an increase in the cost and weight of the electronics installed in the device to be charged. This is represented in Fig. 5.9.
- **Primary- and secondary-side controlled charger**. Both parts of the charger rely on the controller, which modifies their performance according to a number of instantaneous measurements. Their operation may be independent or configured jointly. This type is presented in Fig. 5.10.

The power converters for the three main control approaches will be described next.

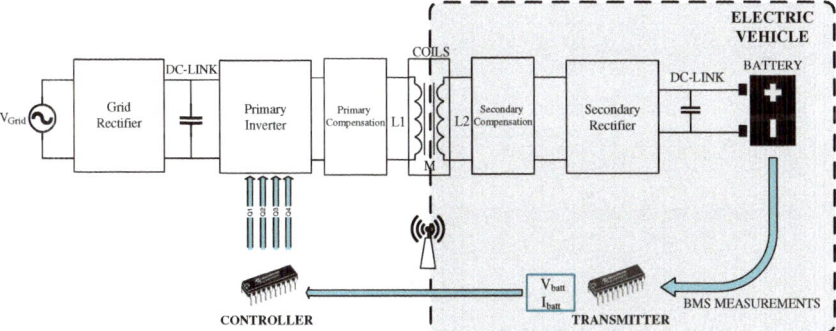

Fig. 5.8 Diagram of a primary-side controlled wireless charger

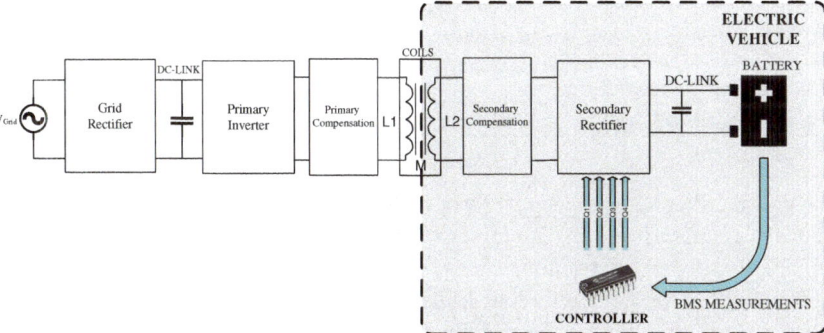

Fig. 5.9 Diagram of a secondary-side controlled wireless charger

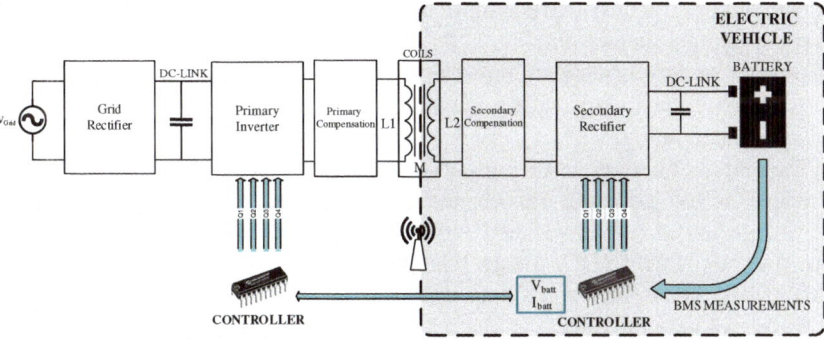

Fig. 5.10 Diagram of a primary and secondary controlled wireless charger

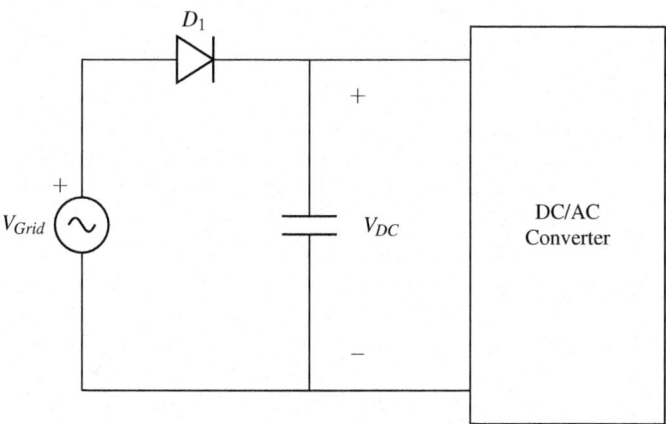

Fig. 5.11 Topology of a half-bridge rectifier

5.3.1 Grid Rectifier

The grid rectifier takes the signal from the grid and generates a DC voltage, which feeds the primary inverter. The most popular rectifiers are composed of a diode or pairs of diodes and a capacitor.

The simplest topology is the half-bridge rectifier, illustrated in Fig. 5.11. The diode acts as a short-circuit when the grid voltage is positive. Once the current provided by the grid becomes negative, the diode turns off. The output signal of the half-bridge is shown in Fig. 5.13. As can be observed, only one half-cycle is transferred to the output voltage. The alternative to make use of both cycles is the full-bridge rectifier, whose structure is depicted in Fig. 5.12. The diodes $D1$ and $D4$ are turned on during the positive semicycle whereas the diodes $D2$ and $D3$ are deactivated for this period. The opposite behaviour occurs during the negative semi-cycle, leading to the output voltage plotted in Fig. 5.13.

The output of both types of rectifiers is connected to a parallel capacitor in order to reduce the voltage ripple. This element constitutes the DC-link. Figure 5.14 shows the voltage and currents associated with a full-bridge rectifier. The output of the DC link is approximately a DC voltage. It is important to note that there is current peak, which will generate EMI and reduce the power factor.

So far, we have analysed one-phase rectifiers, which are appropriate for wireless chargers up to 3.7 kW. For higher power, three-phase rectifiers become necessary instead. There are half-bridge and full-bridge three-phase rectifiers. Their diagrams are presented in Figs. 5.15 and 5.16 respectively.

In the three-phase half-bridge rectifier, the power converter is connected to the grid with a Y-connection. One of the three diodes is always turned on so that the current harmonics are reduced. In the full-bridge three-phase rectifier, there is always a pair

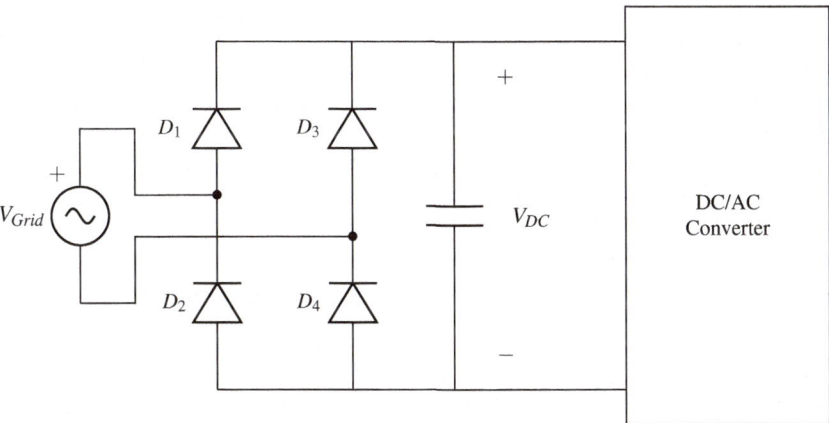

Fig. 5.12 Topology of a full-bridge rectifier

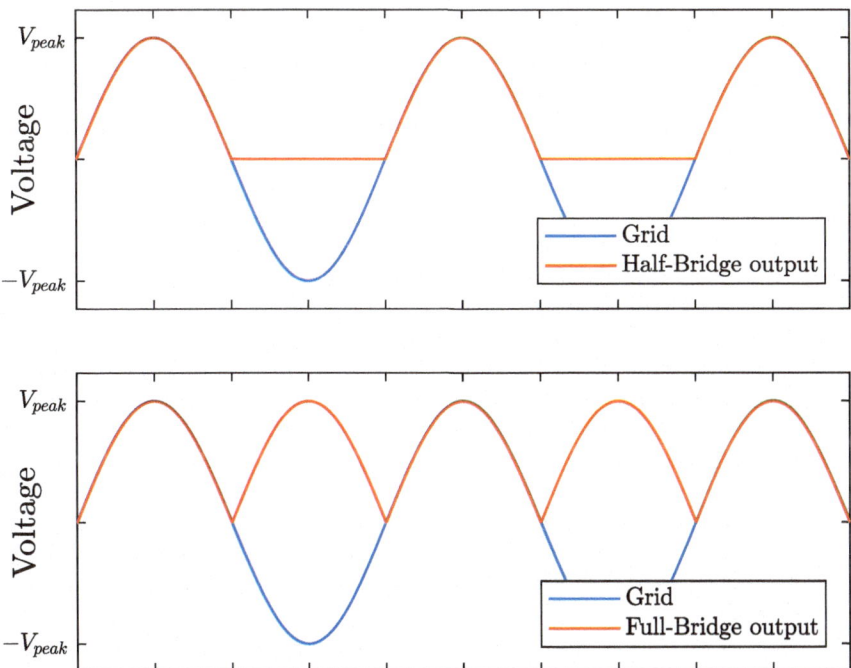

Fig. 5.13 Comparison between the output of a half-bridge rectifier and a full-bridge rectifier

of active diodes, so the ripple in both the output current and voltage is decreased (Fig. 5.17).

Controlled rectifiers substitute the diodes with controllable devices such as thyristors. The activation of the thyristors can be postponed with respect to the voltage

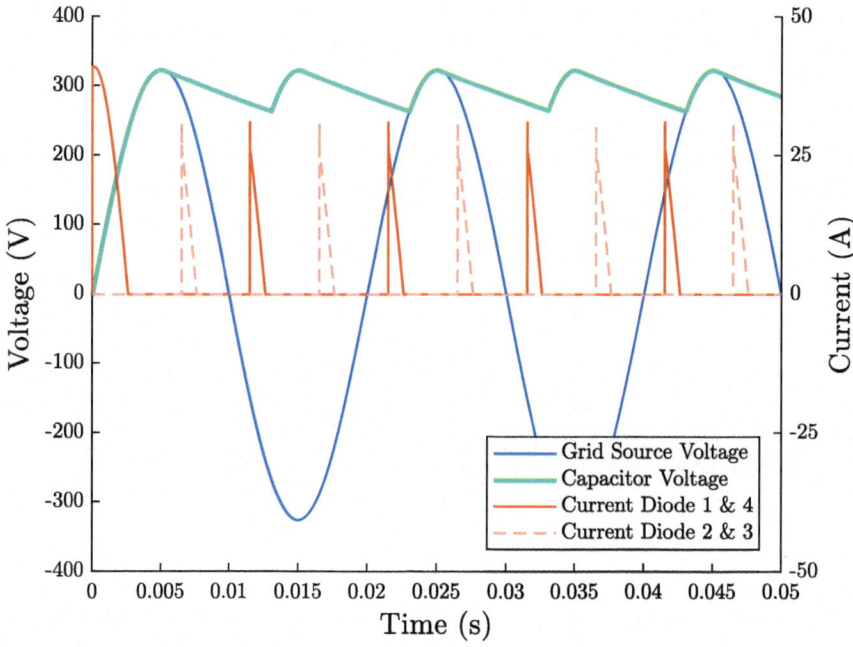

Fig. 5.14 Full-bridge rectifier currents and voltages

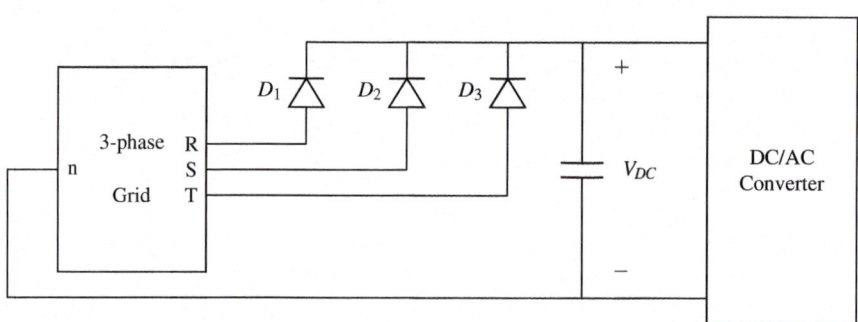

Fig. 5.15 Topology of a 3-phase half-bridge rectifier

activation of the diodes. This control can be used to reduce the harmonics and increase the power factor. The diagram in Fig. 5.18 corresponds to a controllable rectifier. The output current and voltage are depicted in Fig. 5.19.

Power Factor Corrector (PFC) systems may be inserted on the primary side to overcome the problems originated by the rectifier current peaks. These systems can be classified into two main groups:

- **Passive PFC**. This is composed of passive elements to configure an LC filter to attenuate the harmonic currents.

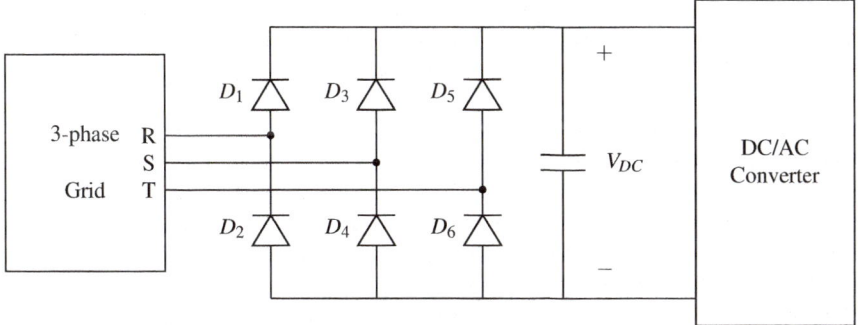

Fig. 5.16 Topology of a 3-phase full-bridge rectifier

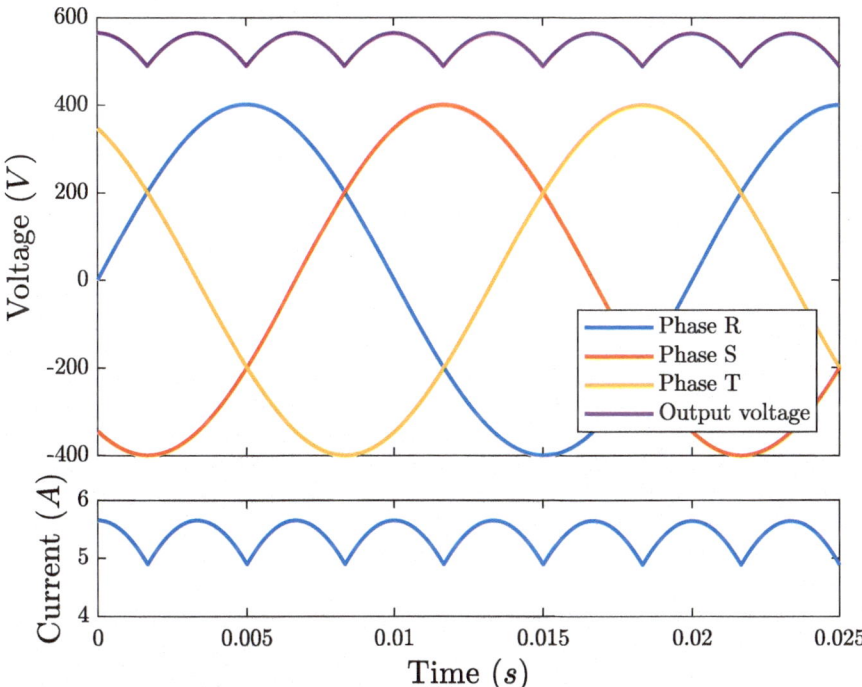

Fig. 5.17 3-phase full-bridge rectifier currents and voltages with a load of 100 Ω

- **Active PFC**. This relies on an additional power converter to modify the grid current appropriately. The components are of smaller size and cheaper than those used in the passive PFC. It can be implemented in a two-stage or a single-stage approach.

The two-stage PFC consists in inserting an independent block and placing it between the primary rectifier and the primary inverter. Its structure is illustrated in Fig. 5.20b. In this case, the usual implementation in wireless chargers is a boost

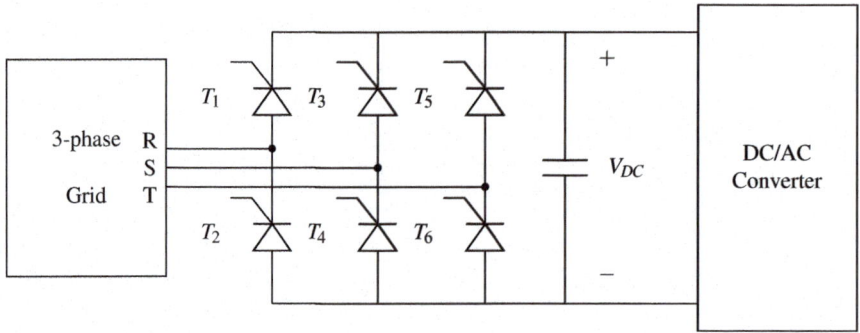

Fig. 5.18 Topology of a 3-phase controlled rectifier

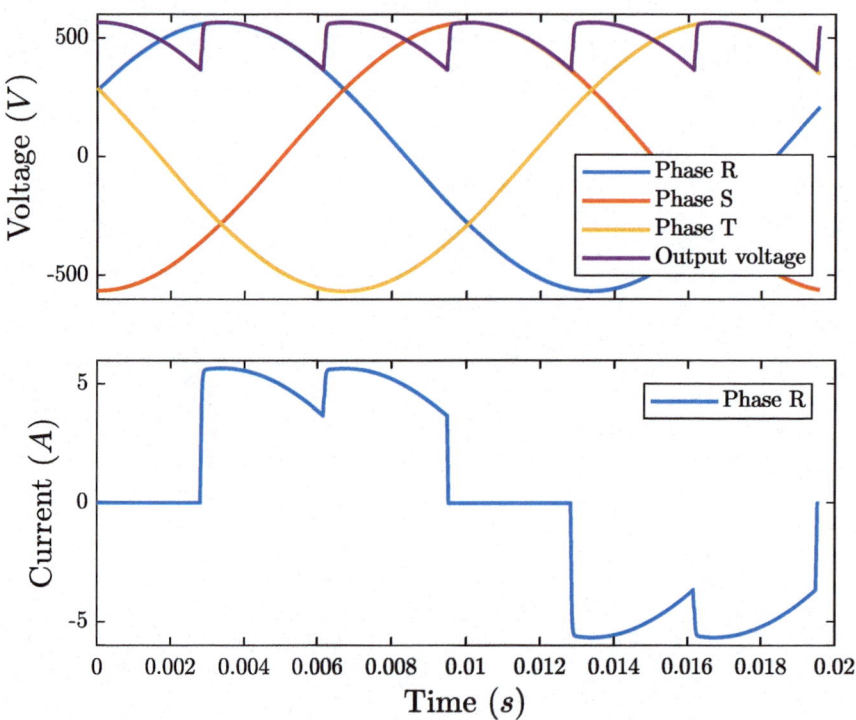

Fig. 5.19 3-phase controlled rectifier currents and voltages with a load of 100 Ω

DC/DC converter [6, 7]. This type of converter increases the output voltage in comparison with the input. An alternative two-stage approach is presented in [29], with a Single-Ended Primary Inductor Converter (SEPIC) structure. This DC-DC converter is known to force its switch to support higher voltages than boost converters for the same power outputs. Although this stress is a clear drawback, SEPIC allows for a complete regulation of the output voltage, i.e. the input voltage can be increased or

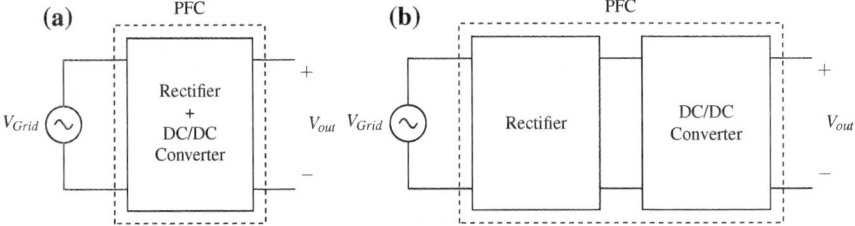

Fig. 5.20 Generic structure of **a** single-stage and **b** two-stage PFC topologies

lowered. In order to cope with this stress feature, the authors in [26] propose the use of two inter-leaved SEPICs as a PFC. Thus, the control is more complex and the application is still restricted to a 500 W prototype.

Including a PFC also minimises harmonic distortion, which has been reported in [16, 17]. The authors of these works have checked this relationship by including a simple and independent PFC in their wireless charger. In particular, their design relies on a single-ended inverter operating with a Pulse Width Modulation (PWM) control.

Alternatively, it is also possible to integrate the PFC in the primary rectifier leading to a single-stage PFC, presented in Fig. 5.20a. In contrast to these previous works, [27] opt for an integrated PFC based on a bridgeless structure. A single-stage PFC is also employed in [4].

A different approach for building a PFC is presented in [7]. Based on a Z-source network, the authors simultaneously control the output power and the power factor delivered by the front-stage without the need to use additional switches. In fact, a Z-source network is built based exclusively on reactive components with no control system applied to them. This makes the proposed PFC more reliable. Its control is indirectly related to the inverter and its duty cycles.

On the other hand, the work in [14] addresses the power factor corrector from a different point of view. The authors in this work propose the design of a LCL compensation topology to control the reactive power delivered back on the utility. The study is theoretical and there is no evidence of how this strategy behaves in a real prototype where the designed values are different from the values of the real components due to tolerances.

5.3.2 Primary Inverter

The inverter DC/AC converter is responsible for generating the high frequency signal involved in the magnetic field. The semiconductor devices in this power converter must switch at a high frequency, a condition that must be considered as they will impact on the losses and on the need for dissipation systems.

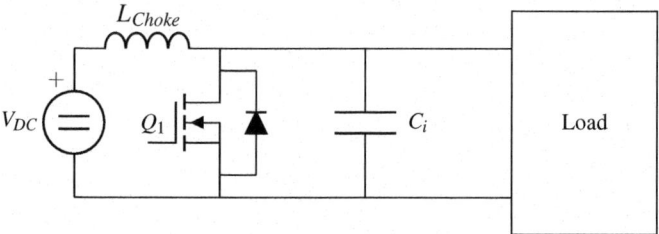

Fig. 5.21 Class-E inverter topology

The output voltage can be controlled to adjust to the battery demands. Through this adjustment, the control system can be restricted to the primary side, which restricts the electronics, costs and weight of the secondary side. Despite these benefits, the control must operate with the switching frequency, which is not a trivial requirement. In contrast to grid-connected inverters, the inverter in a wireless charger may generate a non-sinusoidal signal. A square signal is possible at the output inverter, where the current is modified due to the resonant tanks.

There are multiple configurations for inverters in EV wireless chargers. The simplest topology is the **class-E inverter**, which is illustrated in Fig. 5.21. Only one transistor (usually a MOSFET) is used to switch. There is a DC-feed inductance L_{choke} and a shunt capacitor C_i. The output voltage is a half-cycle of a sinusoidal wave. The frequency of the voltage equals the MOSFET switching. The proper selection of the inductance and the capacitance can lead to Zero-Voltage Source (ZVS) or a Zero-Current Source (ZCS) operation [1, 22]. Other topologies with only one transistor have also been proposed for the inverter. We cite the Second harmonic class E inverter [23] and the Class $\phi 2$ inverter [31]. These kinds of configurations are only feasible for low-power applications as the switch must support all the transferred power [32].

The **Half-Bridge Inverter** or class-D inverter is also a simple solution. The structure is presented in Fig. 5.22. It is composed of two transistors, leading to an output voltage as illustrated in Fig. 5.23. It has been used for low-power WPT systems [18].

The analysis of the output voltage yields to the following expression:

$$V_{out}(t) = D \cdot V_{in} + \sum_{n=1}^{\infty} \frac{2V_{in}}{n\pi} \sin\left(n\pi D\right) \cos\left(n\omega t\right) \tag{5.2}$$

where $\omega = 2\pi f$ and D is the duty cycle. The first harmonic of this wave corresponds to Eq. 5.3.

$$V = \frac{\sqrt{2}}{\pi} V_{DC} \tag{5.3}$$

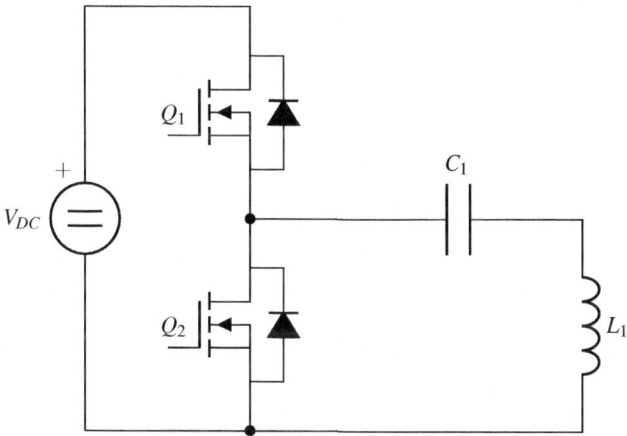

Fig. 5.22 Schematic of a half-bridge inverter with series compensation

The **class-DE inverter** is an intermediate option between the class-D and the class-E inverters [32]. It allows for ZVS switching, which causes a reduction in the electrical stress of the semiconductors. A similar operation is achieved with the Current-fed half-bridge inverter with CLC compensation [24].

For higher power transmissions, the **Full-Bridge Inverter** is the most common implementation. The structure is shown in Fig. 5.24. As can be observed, four transistors are configured in two legs. Drivers are usually incorporated into the systems to ensure the correct activation of the transistors. For each leg the driver is configured to activate a maximum of one transistor. Taking this condition into account, the output voltage depends on the pair of transistors that are activated, so that the potential states are:

- Q1 and Q2 activated. It occurs during the interval $0 \leq t \leq T/2$. The current flows through the load from the left to the right leg and V_{out} is V_{DC}.
- Q3 and Q4 activated. In the interval $T/2 \leq t \leq T$, the current flows in the opposite direction. Consequently, V_{out} is $-V_{DC}$.

The control algorithm will decide the sequence of the states in the full-bridge inverter. For a square-wave control, there are only two states (Q1 and Q2 activated in the first state and Q3 and Q4 activated in the second state). Their duration is the same and equals half of the period. Figure 5.25 depicts the voltage of a full-bridge inverter with a square-wave control. Unlike with the previous inverter topologies, the output ranges from V_{DC} to $-V_{DC}$. In a WPT system, as the circuit usually operates at resonance, the current is filtered to its first harmonic.

Phase-shifting control is a common method to regulate the power levels in a WPT system. With this technique, the output voltage has three potential levels: V_{DC}, $-V_{DC}$ and 0. This latter voltage is produced by the activation of transistors Q1 and Q3 or of the pair Q2 and Q4. Figure 5.26 shows an illustrative example of this technique, in which four states can be distinguished:

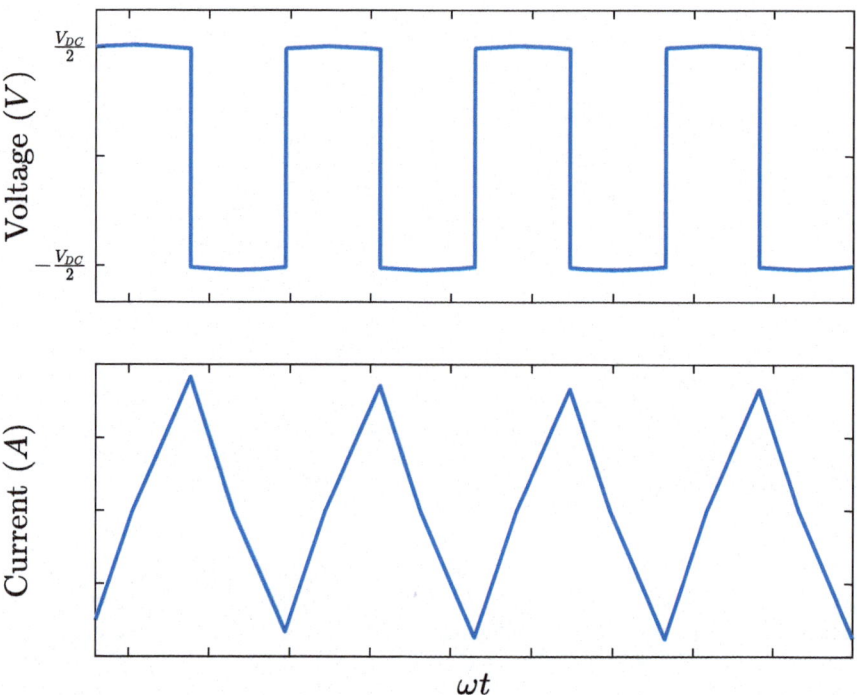

Fig. 5.23 Half-bridge inverter currents and voltages

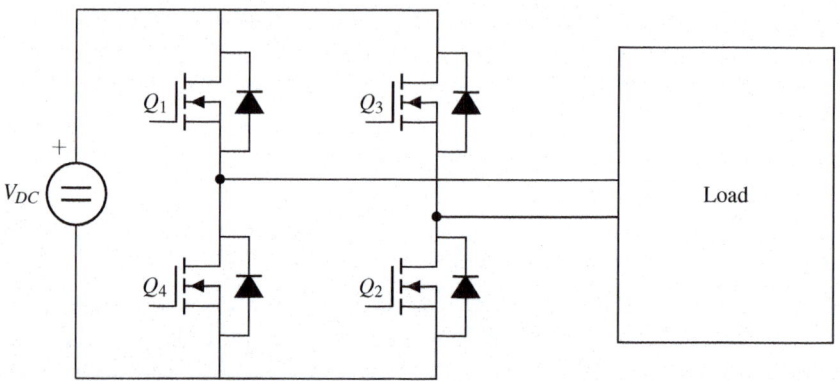

Fig. 5.24 Topology of a full-bridge inverter

- Q1 and Q2 activated. In the interval $0 \leq t \leq D$, the current flows through the load from the left to the right leg and V_{out} is V_{in}.
- Q1 and Q3 activated. It occurs during the interval $D \leq t \leq D + Delay$. Q2 is deactivated, while Q3 is activated. As a consequence, both poles of the load are connected to the positive side and the output voltage is zero.

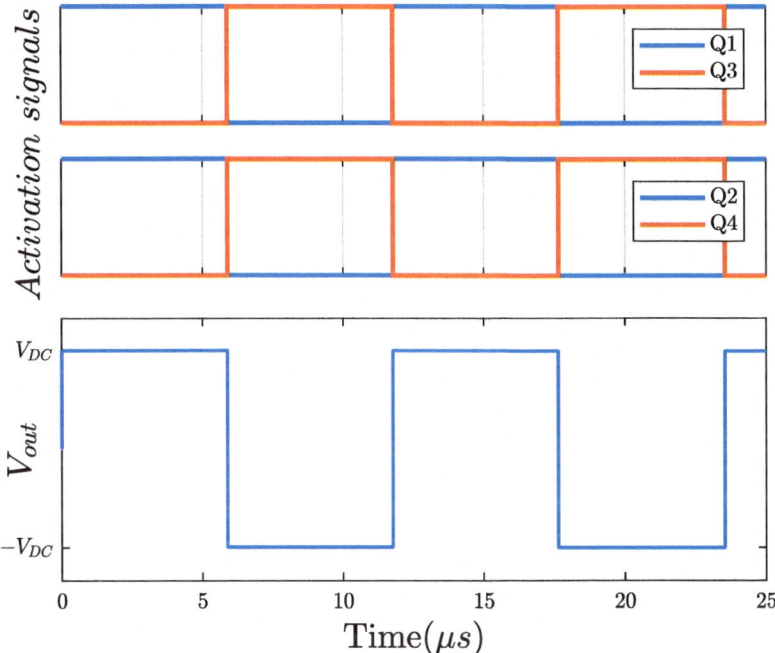

Fig. 5.25 Voltages in a full-bridge inverter with an square-wave activation

- Q3 and Q4 activated. In the interval $D + Delay \leq t \leq T - Delay$, the current flows in the opposite direction. Consequently, V_{out} is $-V_{in}$.
- Q2 and Q4 activated. It occurs during the interval $T - Delay \leq t \leq T$. As when Q1 and Q3 are activated, both poles of the load are connected at the same voltage, so V_{out} is zero.

The analysis of the output voltage yields to the following expression:

$$V_{out}(t) = \sum_{n=1}^{\infty} \frac{4V_{in}}{n\pi} \sin(n\pi D) \sin(n\omega t) \qquad (5.4)$$

As can be observed, the first harmonic is reduced according to parameter $\sin(n\pi D)$. The maximum amplitude for every harmonic component is $4V_{in}/n\pi$ and it is reached when this parameter is equal to 1.

Multi-level inverters are also a solution for the DC/AC converter in high-power applications [2]. It is composed of multiple full-bridge inverters so that the power supported by each switching device is diminished. However, the control algorithm becomes more complex [3].

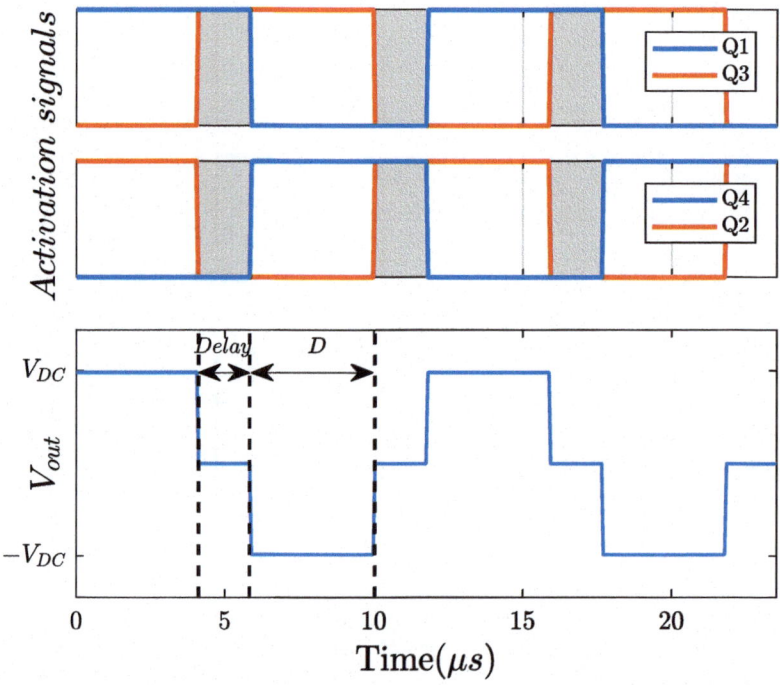

Fig. 5.26 Voltages in a full-bridge inverter with an square-wave activation and phase-shifting

5.3.3 Secondary Rectifier

The most common implementation for this power converter is a non-controlled rectifier, similar to the grid rectifier. The main difference lies in the switching frequency of the devices as they have to support the high frequency imposed by the magnetic field involved in EV wireless charging. The rectifier is connected in parallel with a capacitor to deliver a constant voltage to the battery.

The effects of the secondary rectifier in the AC analysis are commonly modelled as a variation on the battery resistance [13, 28]. The rectifier is a non-linear device which, with its filter, acts as an impedance transformer. The equivalent resistance is computed as presented in Eq. 5.5.

$$R_{eq} = \frac{V_{ac,rms}}{I_{ac,rms}} = \frac{\pi^2}{8} \frac{V_L}{I_L} = \frac{8}{\pi^2} R_L \qquad (5.5)$$

Some solutions opt for including a PFC system on the secondary side. This is the case for the works presented in [9, 14].

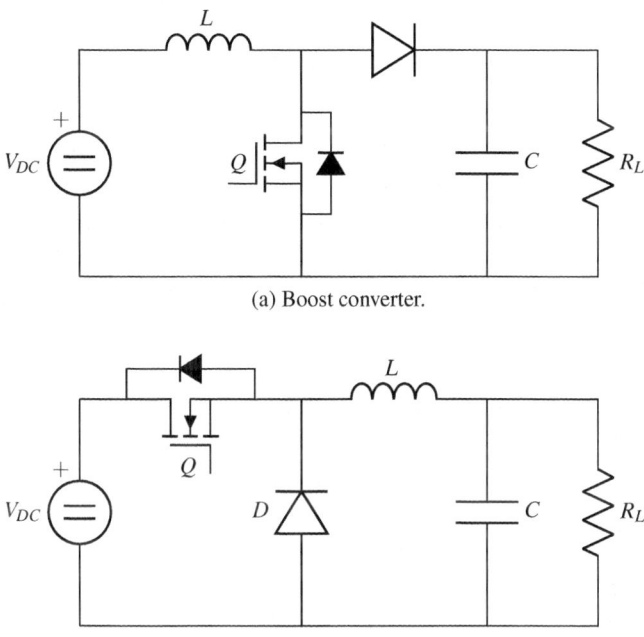

(a) Boost converter.

(b) Buck converter.

Fig. 5.27 DC/DC converter topologies

5.3.4 Secondary DC/DC Converter

This is an optional system inserted between the secondary rectifier and the battery. It adapts the voltage and/or current to the battery requirements. The boost converter increases the output voltage whereas the buck decreases this magnitude. The topologies of both converters are presented in Fig. 5.27.

The output voltage of the boost converter equals to the expression in Eq. 5.6 when the coil is operated with the continuous mode. This means that the current is not null all the time. As can be observed, the output depends on the duty cycle D. The duty cycle represents the percentage of the time period in which the switch is active.

$$V_{out} = \frac{1}{1 - D} V_{in} \qquad (5.6)$$

For discontinuous mode operation, in which the current flowing through the inductor is equal to zero in some periods, the output voltage corresponds to the expression in Eq. 5.7. The proper selection of the coil could force the operation in one of these modes.

$$V_{out} = V_{in} + \frac{V_{in}^2 D^2 T}{2 L I_0} \qquad (5.7)$$

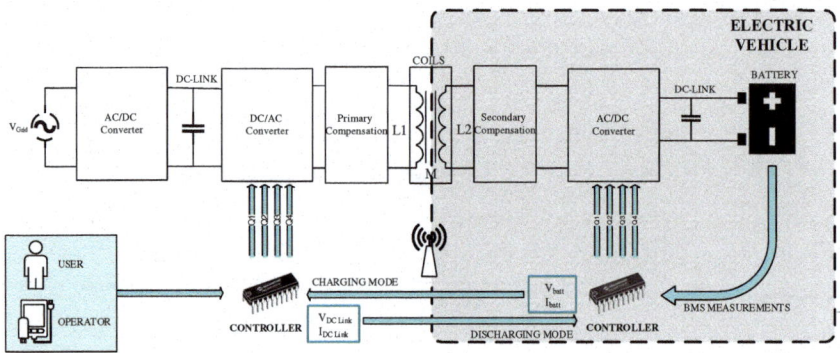

Fig. 5.28 Diagram of a bi-directional wireless charger

In the buck converter, for the continuous mode, the output voltage can be expressed as:

$$V_{out} = \frac{1}{T_s} \int_0^{T_s} v_0(t) \, dt = \frac{1}{T_s} \left(\int_0^{t_{on}} V_{in} \, dt + \int_{t_{on}}^{T_s} 0 \, dt \right) = \frac{t_{on}}{T_s} V_{in} = D V_{in} \quad (5.8)$$

Alternatively, for a discontinuous mode operation, the output voltage is:

$$V_{out} = V_{in} \frac{1}{\frac{2L \cdot I_0}{D^2 \cdot V_{in} \cdot T} + 1} \quad (5.9)$$

5.4   Bi-directional Charger

Bi-directional chargers allow the power to flow from the grid to the battery and vice versa in order to comply with V2G services. To enable power to flow in both directions, power converters must be bi-directional, performing an inverse task in one direction with respect to the other. The control systems configure the direction of the power flow, taking into account certain commands generated by the grid and the restrictions set by users. Figure 5.28 shows a diagram of a bi-directional wireless charger.

5.4.1   Primary AC/DC Converter

In bi-directional chargers, this converter must be able to operate with power flow in both directions. In addition, the integration of the charger into the grid forces two main requirements. Firstly, the harmonics of the signal inserted into the grid

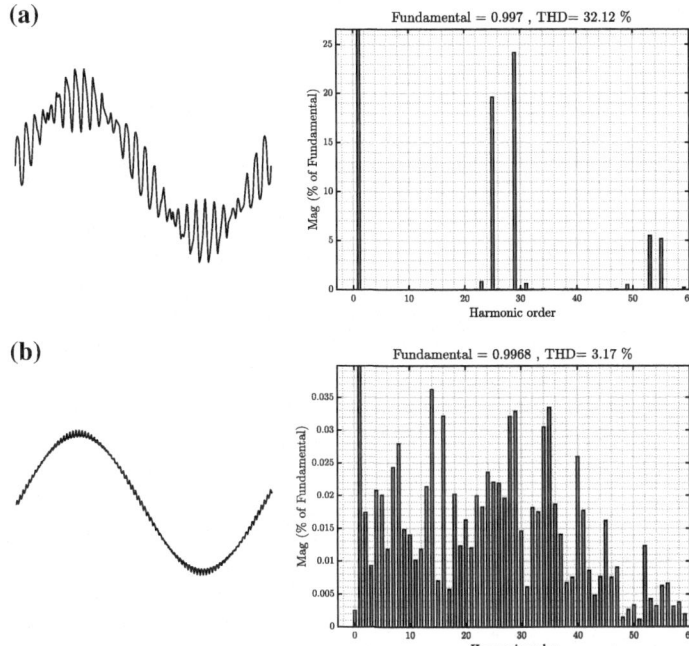

Fig. 5.29 Output signals of a PWM inverter with **a** 1,620 Hz and **b** 4,860 Hz of carrier frequency

must be restricted, as required by IEC [12] and IEEE [11]. As the power converters operate by making their switches commutate, the contents of harmonics in this kind of converter is significant. In fact, the Total Harmonic Distortion (THD), which is a metric showing the impact of harmonics, depends on the topology of the power converter and the control applied to it. PWM or advanced control techniques may be applied to this converter [8, 10, 15, 34]. Figure 5.29 illustrates two examples of the output signal of a PWM inverter with different carrier frequencies and their harmonic content.

In order to integrate the charger into the grid, the harmonics content must be reduced. For this purpose, filters are inserted into the charger, as depicted in Fig. 5.30. They can be composed of purely passive components or rely on configurable power devices. The basic function of their performance is to keep the first harmonic while diminishing the power of the other harmonics.

Secondly, the integration of the converter into the grid requires the signal delivered by the charger to be synchronised with the grid. This requirement is set for wired and wireless chargers, so that the technique by which this converter is configured in a conductive charger can also be applied in wireless systems. The techniques are also implemented for renewable energy sources that provide power to the grid.

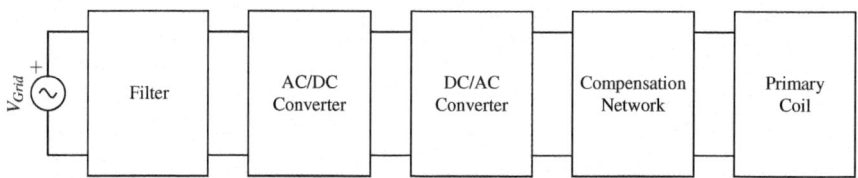

Fig. 5.30 Example of a bidirectional charger's primary side with filter in the grid connection

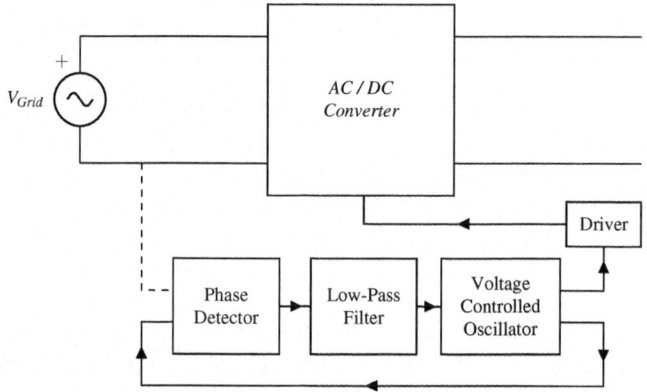

Fig. 5.31 Basic PLL configuration

The simplest method of synchronisation is based on the Zero Crossing Detector (ZCD), which studies the grid signal and detects a transition from positive to negative and vice versa in the sine waveform generated by the grid. With this information, the power converter is controlled. However, ZCD suffers from a high degree of imprecision as it is affected by noise. Phase-Locked Loop (PLL) is a preferable option. PLL measures the grid voltage and obtains the frequency and phase of this signal. Based on these two parameters, the control of the primary AC/DC power converter is dynamically adapted to provide a signal synchronised with the grid. Figure 5.31 shows the diagram of a PLL. The phase detector compares the grid signal with the output of the Voltage-controlled oscillator so that the grid inverter generates a signal that matches that of the grid. The second order generalised integrator phase-locked loop, proposed in [5], outperforms the PLL where there are notable variations in the signal frequency.

A DQ current controller is used when it is necessary to control the active and reactive power provided to the grid. This is the case of EV wireless chargers intended to provide this kind of ancillary service. The generic diagram is depicted in Fig. 5.32.

For high-power applications, the grid converter must provide a three-phase signal. The method by which this converter is configured depends on its topology. For instance, the full-bridge AC/DC converter in Fig. 5.33 may be configured with a bipolar PWM control. The output voltage is presented in Fig. 5.34 together with the carriers for the three phases.

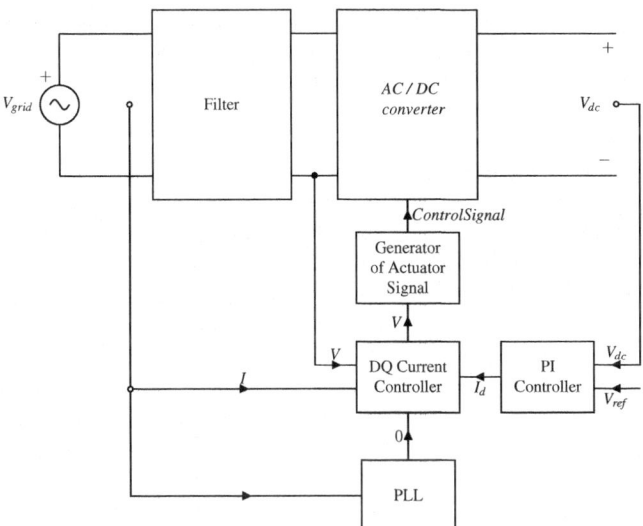

Fig. 5.32 Grid inverter DQ controller

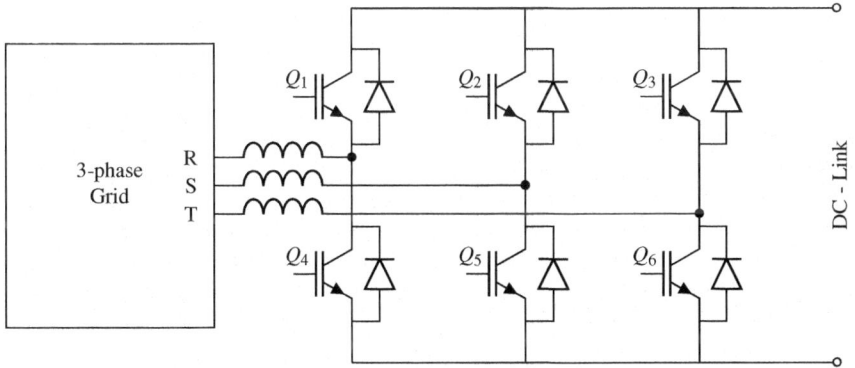

Fig. 5.33 Topology of a three-phase AC/DC converter

Control in a three-phase converter presents a greater challenge than in the mono-phase systems as there are more signals to control. Additional computation and signal samples are required. For three-phase DC/AC converters, the dq control can be used to adjust the active and reactive power delivered to the grid, as used in single-phase inverters. The control based on the *abc* transformation allows the current in each phase to be controlled independently.

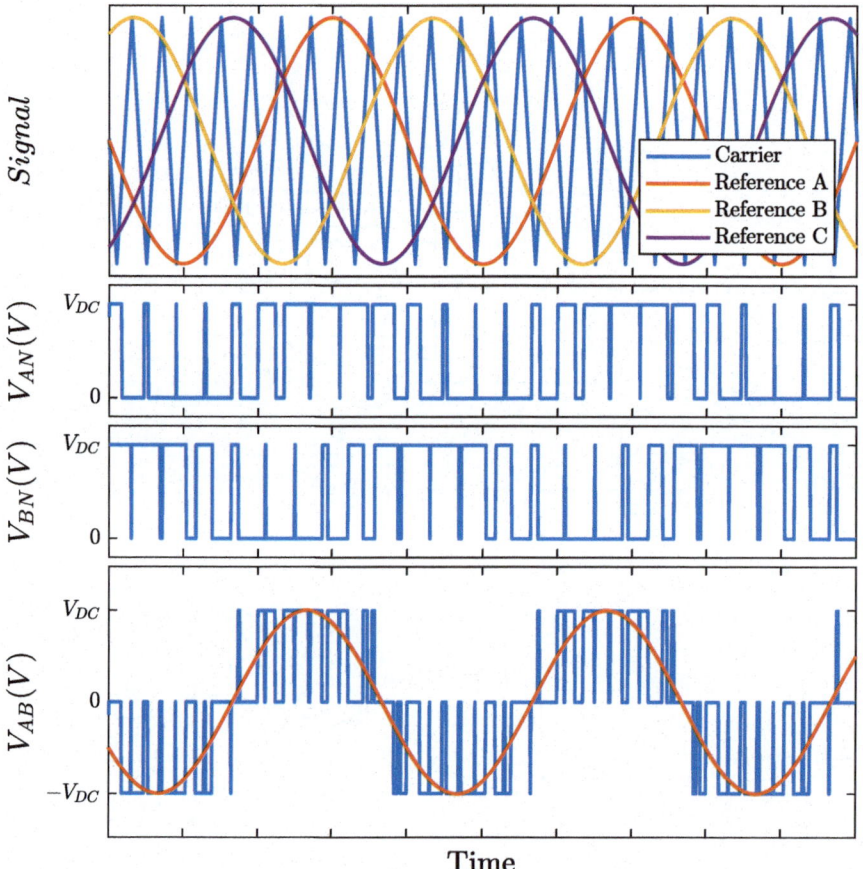

Fig. 5.34 Three-phase inverter output

5.4.2 Secondary AC/DC Converter

This power converter must allow the power flow in both directions. Controlled rectifiers can be configured as inverters if the firing angles range from 90° to 180°. The uni-directional inverters (e.g. the one illustrated in Fig. 5.24) can be modified to be bi-directional if we ensure that the flywheel diodes support a higher intensity of current.

5.4.3 Secondary DC/DC Converter

These types of converters are optional in wireless chargers as they are inserted only when there is a difference in the voltage provided by the secondary AC/DC converter and the battery voltage.

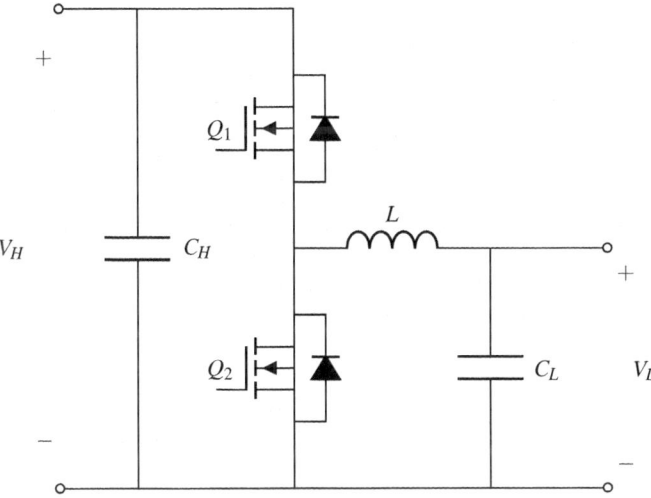

Fig. 5.35 Bidirectional DC/DC converter: Type-C chopper

There are some converters that allow an increment in the voltage in one direction and a decrement in the another. This is achieved by the use of more than one switch. A type-C chopper, or two-quadrant chopper, relies on two switches for this function, with only one being operative in each direction of the power flow. The diagram of this converter is shown in Fig. 5.35. When switch Q1 is controlled, the converter behaves similarly to a buck converter. Alternatively, if it is the switch Q2 that is controlled, the converter acts as a boost converter. Only one coil is required, therefore a Type-C chopper is simple to implement.

Some advanced proposals include isolation of the bi-directional DC-DC converters as in [20, 30]. However, these solutions are not useful in wireless chargers as they already have isolation due to the air-cored coupled coils.

References

1. Aldhaher, S., Luk, P.C.K., Whidborne, J.F.: Electronic Tuning of Misaligned Coils in Wireless Power Transfer Systems. IEEE Trans. Power Electron. **29**(11), 5975–5982 (2014). https://doi.org/10.1109/TPEL.2014.2297993, http://ieeexplore.ieee.org/document/6702433/
2. Asa, E., Colak, K., Bojarski, M., Czarkowski, D.: A novel multi-level phase-controlled resonant inverter with common mode capacitor for wireless EV chargers. In: 2015 IEEE Transportation Electrification Conference and Expo (ITEC), pp. 1–6. IEEE (2015). https://doi.org/10.1109/ITEC.2015.7165734, http://ieeexplore.ieee.org/document/7165734/
3. Bojarski, M., Asa, E., Outeiro, M.T., Czarkowski, D.: Control and analysis of multi-level type multi-phase resonant converter for wireless EV charging. In: IECON 2015 - 41st Annual Conference of the IEEE Industrial Electronics Society, pp. 005,008–005,013. IEEE (2015). https://doi.org/10.1109/IECON.2015.7392886, http://ieeexplore.ieee.org/document/7392886/

4. Chinthavali, M., Onar, O.C., Campbell, S.L., Tolbert, L.M.: Isolated wired and wireless battery charger with integrated boost converter for PEV applications. In: 2015 IEEE Energy Conversion Congress and Exposition (ECCE), pp. 607–614. IEEE (2015). https://doi.org/10.1109/ECCE. 2015.7309745, http://ieeexplore.ieee.org/document/7309745/

5. Ciobotaru, M., Teodorescu, R., Blaabjerg, F.: A new single-phase pll structure based on second order generalized integrator. In: 37th IEEE Power Electronics Specialists Conference, pp. 1–6. IEEE. 10.1109/PESC.2006.1711988. http://ieeexplore.ieee.org/document/1711988/

6. Deng, J., Fei Lu, Li, S., Nguyen, T.D., Mi, C.: Development of a high efficiency primary side controlled 7kW wireless power charger. In: 2014 IEEE International Electric Vehicle Conference (IEVC), pp. 1–6. IEEE (2014). https://doi.org/10.1109/IEVC.2014.7056204, https:// ieeexplore.ieee.org/document/7056204/

7. Gonzalez-Santini, N.S., Zeng, H., Yu, Y., Peng, F.Z.: Z-source resonant converter with power factor correction for wireless power transfer applications. IEEE Trans. Power Electron. 31(11), 7691–7700 (2016). https://doi.org/10.1109/TPEL.2016.2560174, http://ieeexplore.ieee.org/ document/7462271/

8. Ho, C.M., Cheung, V., Chung, H.H.: Constant-frequency hysteresis current control of grid-connected VSI without bandwidth control. IEEE Trans. Power Electron. 24(11), 2484–2495 (2009). https://doi.org/10.1109/TPEL.2009.2031804, http://ieeexplore.ieee.org/ document/5233830/

9. Hsieh, Y.C., Lin, Z.R., Chen, M.C., Hsieh, H.C., Liu, Y.C., Chiu, H.J.: High-efficiency wireless power transfer system for electric vehicle applications. IEEE Trans. Circuits Syst. II: Express Briefs 64(8), 942–946 (2017). https://doi.org/10.1109/TCSII.2016.2624272, http://ieeexplore. ieee.org/document/7731201/

10. Hu, J., Zhu, J., Dorrell, D.G.: Model predictive control of grid-connected inverters for PV systems with flexible power regulation and switching frequency reduction. IEEE Trans. Ind. Appl. 51(1), 587–594 (2015). https://doi.org/10.1109/TIA.2014.2328785, http://ieeexplore. ieee.org/lpdocs/epic03/wrapper.htm?arnumber=6825852

11. IEEE Standards Association: IEEE 519-2014 - IEEE Recommended Practice and Requirements for Harmonic Control in Electric Power Systems. https://standards.ieee.org/standard/ 519-2014.html

12. International Electrotechnical Commission (IEC): EMC Product Family Standards: Emission. https://www.iec.ch/emc/emc_prod/prod_emission.htm

13. Junjun Deng, Siqi Li, Sideng Hu, Mi, C.C., Ruiqing Ma: Design methodology of LLC resonant converters for electric vehicle battery chargers. IEEE Trans. Veh. Technol. 63(4), 1581–1592 (2014). 10.1109/TVT.2013.2287379. http://ieeexplore.ieee.org/document/6648465/

14. Keeling, N., Covic, G., Boys, J.: A unity-power-factor IPT pickup for high-power applications. IEEE Trans. Ind. Electron. 57(2), 744–751 (2010). https://doi.org/10.1109/TIE.2009.2027255, http://ieeexplore.ieee.org/document/5173523/

15. Kennel, R., Linder, A.: Predictive control of inverter supplied electrical drives. In: 2000 IEEE 31st Annual Power Electronics Specialists Conference. Conference Proceedings (Cat. No.00CH37018), vol. 2, pp. 761–766. IEEE. https://doi.org/10.1109/PESC.2000.879911, http://ieeexplore.ieee.org/document/879911/

16. Kitamoto, T., Omori, H., Murakami, A., Morizane, T., Kimura, N., Nakaoka, M.: A novel type of high power-factor miniaturized wireless ev charger with optimized power receiving circuit and single-ended inverter. In: 2016 IEEE International Power Electronics and Motion Control Conference (PEMC), pp. 248–253. IEEE (2016). https://doi.org/10.1109/EPEPEMC. 2016.7752006, http://ieeexplore.ieee.org/document/7752006/

17. Kitano, Y., Omori, H., Kimura, N., morizane, T., Nakagawa, K., Nakaoka, M.: A new wireless EV charger using single switch ZVS resonant inverter with optimized power transfer and low-cost PFC. In: 2015 International Conference on Electrical Drives and Power Electronics (EDPE), pp. 515–521. IEEE (2015). https://doi.org/10.1109/EDPE.2015.7325347, http:// ieeexplore.ieee.org/document/7325347/

18. Li, H., Li, J., Wang, K., Chen, W., Yang, X.: A maximum efficiency point tracking control scheme for wireless power transfer systems using magnetic resonant coupling.

IEEE Trans. Power Electron. **30**(7), 3998–4008 (2015). https://doi.org/10.1109/TPEL.2014.
2349534, http://ieeexplore.ieee.org/document/6880373/

19. Lin, H., Villamor, A.: Power SiC 2018: Materials, devices and applications. Technical report, Yole Developpement (2018). https://www.i-micronews.com/report/product/power-sic-2018-materials-devices-and-applications.html#companies-cited

20. Ma, G., Wenlong, Q., Gang, Y., Liu, Y., Liang, N., Li, W.: A zero-voltage-switching bidirectional DCDC converter with state analysis and soft-switching-oriented design consideration. IEEE Trans. Ind. Electron. **56**(6), 2174–2184 (2009). https://doi.org/10.1109/TIE.2009.2017566, http://ieeexplore.ieee.org/document/4808150/

21. McBryde, J., Kadavelugu, A., Compton, B., Bhattacharya, S., Das, M., Agarwal, A.: Performance comparison of 1200V Silicon and SiC devices for UPS application. In: IECON 2010 - 36th Annual Conference on IEEE Industrial Electronics Society, pp. 2657–2662. IEEE (2010). https://doi.org/10.1109/IECON.2010.5675125, http://ieeexplore.ieee.org/document/5675125/

22. Niefnecker, P., Simon, M., Salich, S., Pforr, J.: Comparison of switching devices for a zero-current switched class E based automotive inductive charging converter system. In: 2017 19th European Conference on Power Electronics and Applications (EPE'17 ECCE Europe), pp. P.1–P.10. IEEE (2017). https://doi.org/10.23919/EPE17ECCEEurope.2017.8099265, http://ieeexplore.ieee.org/document/8099265/

23. Rivas, J.M., Han, Y., Leitermann, O., Sagneri, A.D., Perreault, D.J.: A high-frequency resonant inverter topology with low-voltage stress. IEEE Trans. Power Electron. **23**(4), 1759–1771 (2008). https://doi.org/10.1109/TPEL.2008.924616, http://ieeexplore.ieee.org/document/4558255/

24. Samanta, S., Rathore, A.K.: A new inductive wireless power transfer topology using current-fed half bridge CLC transmitter LC receiver configuration. In: 2016 IEEE Energy Conversion Congress and Exposition (ECCE), pp. 1–8. IEEE (2016). https://doi.org/10.1109/ECCE.2016.7854716, http://ieeexplore.ieee.org/document/7854716/

25. She, X., Huang, A.Q., Lucia, O., Ozpineci, B.: Review of silicon carbide power devices and their applications. IEEE Trans. Ind. Electron. **64**(10), 8193–8205 (2017). https://doi.org/10.1109/TIE.2017.2652401, http://ieeexplore.ieee.org/document/7815312/

26. Shi, C., Khaligh, A., Wang, H.: Interleaved SEPIC power factor preregulator using coupled inductors in discontinuous conduction mode with wide output voltage. IEEE Trans. Ind. Appl. **52**(4), 3461–3471 (2016). https://doi.org/10.1109/TIA.2016.2553650, http://ieeexplore.ieee.org/document/7452383/

27. Siu, K.K., Ho, C.N.: A critical review of Bridgeless PFC boost rectifiers with common-mode voltage mitigation. In: IECON 2016 - 42nd Annual Conference of the IEEE Industrial Electronics Society, pp. 3654–3659. IEEE (2016). https://doi.org/10.1109/IECON.2016.7793232, http://ieeexplore.ieee.org/document/7793232/

28. Steigerwald, R.: A comparison of half-bridge resonant converter topologies. IEEE Trans. Power Electron. **3**(2), 174–182 (1988). https://doi.org/10.1109/63.4347, http://ieeexplore.ieee.org/document/4347/

29. STMicroelectronics: AN2435: TM sepic converter in PFC pre-regulator (2007)

30. Texas Instruments: TI Designs: TIDA-BIDIR-400-12 Bidirectional DC-DC Converter (2015). http://www.ti.com/lit/ug/tiduai7/tiduai7.pdf

31. Uddin, M.K., Kalwar, K.A., Mekhilef, S., Ramasamy, G.: A wireless vehicle charging system using Class phi2 inverter. In: 2015 IEEE PELS Workshop on Emerging Technologies: Wireless Power (2015 WoW), pp. 1–5. IEEE (2015). https://doi.org/10.1109/WoW.2015.7132801, http://ieeexplore.ieee.org/document/7132801/

32. Uddin, M.K., Ramasamy, G., Mekhilef, S., Ramar, K., Lau, Y.C.: A review on high frequency resonant inverter technologies for wireless power transfer using magnetic resonance coupling. In: 2014 IEEE Conference on Energy Conversion (CENCON), pp. 412–417. IEEE (2014). https://doi.org/10.1109/CENCON.2014.6967539, http://ieeexplore.ieee.org/document/6967539/

33. Wang, F.F., Zhang, Z.: Overview of silicon carbide technology: device, converter, system, and application. CPSS Trans. Power Electron. Appl. **1**(1), 13–32 (2016). https://doi.org/10.24295/CPSSTPEA.2016.00003, http://tpea.cpss.org.cn/uploads/soft/170214/1_1511396481.pdf
34. Yao, Z., Xiao, L., Yan, Y.: Dual-buck full-bridge inverter with hysteresis current control. IEEE Trans. Ind. Electron. **56**(8), 3153–3160 (2009). https://doi.org/10.1109/TIE.2009.2022072, http://ieeexplore.ieee.org/document/4926184/

Chapter 6
Design Procedure of an EV Magnetic Resonance Charger

6.1 Introduction

Designing a WPT charger is not trivial since it must comply with multiple requirements related to magnetic emissions, efficiency, cost and dimensions.

All these aspects must be considered together, as they are all linked. Based on our experience, we propose the flowchart in Fig. 6.1 as the recommended steps to follow when designing and prototyping an EV wireless charger. This is composed of 8 phases. For the sake of simplicity, the 8 phases are depicted as sequential but some intermediate iterations may be needed when using complex coil structures and advanced power converters.

The first phase, "**Requirement specifications**", consists in defining the main requirements of the WPT application. Among these, the following items stand out as the most relevant:

- Power delivered to the battery. This must be set if the power to transfer is high (greater than 11 kW), medium (from 1 to 11 kW) or low (smaller than 1 kW).
- Nominal voltage at the battery or current required during the charge mode. EV batteries differ in their electrical configurations during the charge time.
- Charging mode, i.e. how the battery parameters must be configured. As explained in Chap. 2, the most common charging modes are constant voltage, constant current, CC–CV or MCC.
- Operation mode, which identifies how the power transfer takes place depending on whether the load is static, stationary or mobile.
- Capacity for bi-directional power flow if V2G services are to be implemented.
- Dimension restrictions of the power transmitter and the power receiver.
- Weight restriction of the power transmitter and receiver. This is a major limitation in UAVs or electric bicycles, for instance.
- Costs. Although the costs are always expected to be minimum, the configuration of the power transmitter and the receivers can be designed specifically to minimise one of these parts. For instance, public institutions may develop more expensive

© Springer Nature Switzerland AG 2020
A. Triviño-Cabrera et al., *Wireless Power Transfer for Electric Vehicles: Foundations and Design Approach*, Power Systems,
https://doi.org/10.1007/978-3-030-26706-3_6

Fig. 6.1 Generic flowchart for the design and implementation of an EV WPT charger

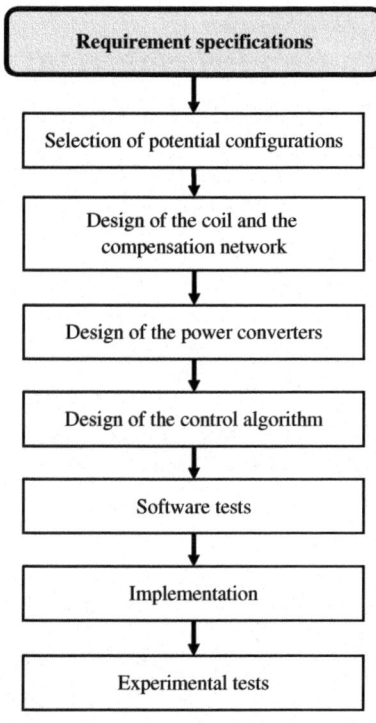

power transmitters for the EV charger if that could reduce the price of the power receivers.

- Ability to cope with misalignment. The degree of misalignment that the load may take should be clearly established in this phase, as it will greatly impact the coil design, the power converters and the decision made regarding the compensation networks.
- Operational frequency. For EV wireless charger complying with the recommended practice SAE J2954 [11], the frequency of 85 kHz is the operational frequency to set. Other applications may opt for other values.
- Control of electromagnetic emissions. Depending on the scenario in which the EV wireless charger is to be applied, it will be necessary to restrict the electromagnetic emissions. In factories with autonomous EVs and a low human presence, this control will be achieved in a different way compared with a public wireless charger in a parking lot.

Once the specifications are defined, we proceed with the selection of the potential configurations. These configurations depend on certain factors: the input voltage, the charging voltage, the position of the charging power control (primary or secondary side), the complexity of the system, the minimum efficiency, etc. For example, if the charging power has to be controlled on the secondary side, it is necessary to use a DC/DC converter on that side. On the other hand, if the charging power can be con-

Table 6.1 Analysis of requirements

Parameter	Options	Main impacts
Power transferred	• High • Medium • Low	• Criterion to use in the design: maximum power transfer or maximum efficiency • Materials of the coils • Power devices • Topologies of the power converters
Battery features	• Nominal current • Nominal voltage	• Power devices in the secondary side • Topologies of the power converters on the primary and the secondary side • Need for a DC/DC converter
Charging mode	• CC • CV • CC–CV • MCC	• Topologies of the compensation networks • Control algorithms • Need for a DC/DC converter
Operation mode	• Static • Stationary • Dynamic	• Coil design • Shielding design
Bi-directional power flow	• No implemented • V2L • V2G	• Topologies of the power converters • Addition of semiconductor devices for the two-way power flow • Complex controllers to ensure grid synchronisation • Need for control algorithms to adjust reactive and/or active power
Cost minimisation	• Global • On the primary side • On the secondary side	• Coil design • Topologies of the power converters • Compensation networks • Control algorithms
Operational frequency	• CC • Fixed or variable • SAE compliant	• Analysis and control of bifurcation if variable frequency • Power devices to support the switching frequency • Controllers with appropriate A/D input to process electric measurements at the operational frequency

trolled on the primary side, the converter is not necessary. Table 6.1 summarises the relationship between the requirement and the design impacts. As can be observed, the amount of power transferred has an impact on the design criterion. For low-power applications, the maximum power transfer is normally used to set the values of the impedance matching networks. In contrast, high-power applications are designed with the aim of achieving high efficiency while reducing the losses in the coupler. This is also the reason why coils are built with Litz wire in high-power applications, whereas copper cable is usual in low-power systems. The selection of the power devices is closely related to the power transferred. The switches in the semiconductors must support a proportion of this power and so must be selected according to this parameter and the topology of the power converters. For instance, low-power applications may opt for half-bridge inverters whereas high-power systems require full-bridge inverters.

Manufacturers offer batteries with different nominal voltage, which greatly differ according to the type of EV. It is possible to find batteries with a voltage up to 800 V [3] for an electric car, whereas an e-bike battery ranges from 33 to 75 V. The batteries are designed to have an optimised charge in one of the charging modes studied in Chap. 2 (CC, CV, CC–CV or MCC). Consequently, the compensation system should be selected to allow the charger to operate as a current or a voltage source, as described in Chap. 4.

The coil design should be based on the mode in which the power transfer will take place. As explained in Chap. 3, coils for dynamic WPT systems differ from those used in static or stationary chargers. Shielding and the geometry of the ferrite materials are also adapted to the coil structure. The design of the coupler is key for the proper performance of the system. It not only affects the efficiency, but also the behaviour under coil misalignments. With no misalignment, circular coils have a higher self-inductance than square coils but they are highly sensitive to misalignment (large variation of the mutual inductance). For EV wireless chargers, square or rectangular coils are more common in order to offer an acceptable efficiency even when misalignment occurs.

The capacity for bi-directional WPT adds complexity to the charger. First, the power converters must be bi-directional. Next, the charger must require the control of the active and reactive power delivered to the grid as specified in some ancillary services. The controller becomes more complex in order to offer this functionality. When the power is delivered directly to the grid, the signal must also be synchronised. For this case, the controllers should have the algorithms to synchronize with the grid.

As a generic approach, the wireless charge must be implemented at a minimum cost while still satisfying all the requirements. This includes both the transmitter and the receiver components. Under some circumstances, the price of the charger may be increased in the transmitter if this guarantees that the receiver is cheaper. This is the case for public or private institutions developing an infrastructure of chargers to be used by multiple users. An attractive way to increase the use of their infrastructure is by reducing the price of the electronics in the receivers. With this approach, the secondary DC/DC converter should be avoided and the control itself may be implemented only on the primary side so that the controller is suppressed in the power receiver.

Another relevant issue to resolve in the first phase is the operational frequency and whether it should be fixed or have potential variations. One method to cope with misalignment consists in varying the operational frequency. If this technique is applied, then the system should be designed to avoid bifurcation and this condition should be included in the coil design process.

Considering the system specifications and their impacts on the design of the charger components, we have to identify the potentially valid configurations that satisfy the requirements in phase 2, "**Selection of potential configurations**". In this phase, we mainly have to determine:

Table 6.2 Inputs and outputs of the algorithm to design the coils and compensation networks

Inputs	Outputs
Valid compensation networks	Primary coil
Potential coil geometries	Secondary coil
Maximum coil size	Type of wire
Need for shielding	Primary compensation network
Minimum efficiency	Secondary compensation network
Bifurcation allowed	Theoretical performance
Range of misalignment	

- The valid topologies of the compensation networks to comply with the charging mode of the battery and the requirement to support misalignment. LCC-topology offers a robust solution to cope with misalignment but it is complex to design and can become costly. Series-Series provides a feasible solution to operate under ZPA for different misalignment conditions.
- The options for the coil geometries and the shielding considering the economic and dimensional restrictions.

The design of the coils is a critical phase. In this phase—"**Design of the coils and the compensation networks**"—an iterative algorithm is executed to evaluate the feasibility and performance of each solution. For each coil, it is necessary to determine: its geometry, the type (size and number of strands) of wire used, the number of turns and the shielding geometry if applied. When deciding these, the performance of the power transfer is evaluated in a circuit equivalent to those used in Chap. 4. This type of circuit models the coupled coils and the compensation networks. The battery's electrical features must be considered when computing the value of R_L. When carrying out the evaluation, the main metrics analysed to characterise the performance of the system are the efficiency, the power delivered to the load and the design of the system with no bifurcation.

The next table summarises the inputs and outputs of the proposed algorithm to jointly design the coils and compensation networks (Table 6.2).

Figure 6.2 shows the proposed iterative algorithm to jointly design the coils and compensation networks. It consists in exploring the complete solution space resulting from the potential compensation networks and the allowed coil geometries so that the performance requirements are met. Once all the valid combinations are identified, they are evaluated according to the performance of the power transfer and one of them is selected based on these metrics. $N1$ and $N2$ correspond to the number of turns of the primary and secondary coils respectively.

The phase "**Design of the power converters**" includes the definition of the topologies of the power converters, the selection of the semiconductor devices and their drivers. The decisions taken in this phase will depend on the power to transfer and the efficient control of this parameter.

Fig. 6.2 Flowchart for the
design of the coils and
compensation networks

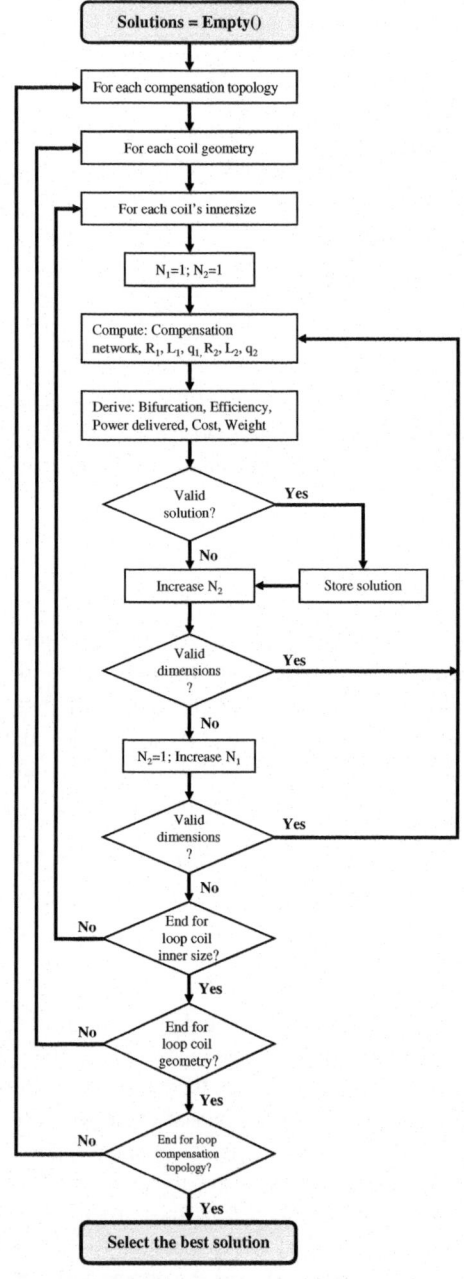

Regarding the "**Design of the control algorithms**", this will rely on the topologies of the power converters. It is necessary to consider the transient periods allowed to adjust the control and the communication systems between the primary and the secondary side. Sampling of the electrical variables must be accomplished taking into account the microprocessor's frequency.

Once the design of the components has been finished, software tests must be carried out. In particular, the phase referred to as "**Software tests**" aims to verify the theoretical performance obtained in the previous phases. Electronics software such as PSIM and Simulink are useful for this task.

The next phases are the "**Implementation**" and the "**Lab experiments**" to check the performance.

6.2 Illustrative Design Procedure

To illustrate how to proceed with the design, implementation and testing of an EV wireless charger, we describe our experience in building a 3.7-kW prototype. We have followed the phases in the previous section, as we will describe next.

6.2.1 Requirement Specifications

Following the specifications defined in the recommended practice SAE J2954 [11], a prototype WPT Power Class 1 and Z-class 2 will be developed during this chapter. The requirements of the prototype are:

- The power delivered to the battery should be 3.7 kW. This charging power is set by the WPT Power Class 1.
- The nominal voltage of the battery is 300 V, which corresponds to a conventional nominal voltage in electric cars [5].
- Static operation mode.
- Dimensional restrictions for the secondary coil. The gap between the primary and the secondary coil could range from 140 to 210 mm, 200 mm being the nominal gap.
- Reduced weight on the secondary side.
- Reduced costs, especially due to the coils' materials.
- Simple control algorithms.
- Capacity to cope with misalignment.
- The operational frequency is 85 kHz. This value is defined in the SAE J2954, although it can fluctuate in the range between 81.38 and 90 kHz.

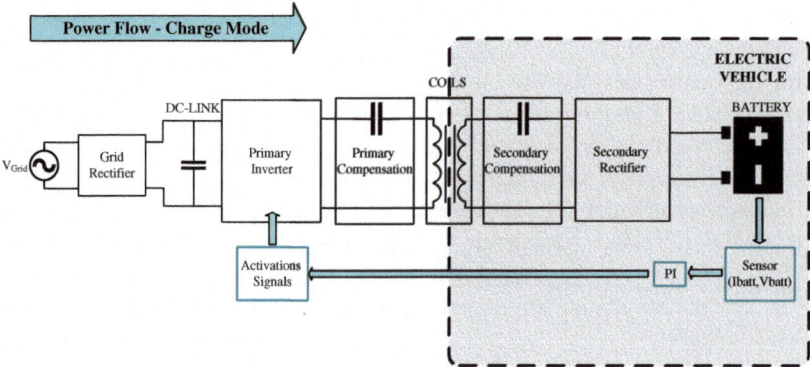

Fig. 6.3 Topology of the illustrative wireless charger

6.2.2   Selection of Potential Configurations

Aimed at developing simple control algorithms, a mono-resonant Series-Series compensation network has been selected. This type of compensation network is less sensitive to misalignment (the correct value of C_1 does not depend on M) than the other mono-resonant topologies. In Series-Series compensation networks, the current flowing through the capacitors is lower than that derived from a parallel connection. This also reduces costs. As the potential coil misalignment ranges over a considerable interval, the design of the EV wireless charger is based on square coils. The structure is depicted in Fig. 6.3.

The structure of the wireless charge is composed of a three-phase rectifier with a capacitor, leading to the DC link. The DC link feeds the inverter to generate the 85-kHz signal. Only a rectifier is included on the secondary side, meaning that no DC–DC converter is used thereby reducing costs and complexity. With this configuration, the charging parameters must be controlled by the primary inverter.

6.2.3   Design of the Coils and Compensation Networks

The process of designing the coils begins by defining their maximum and minimum dimensions. We decided to use a bigger coil on the primary side than on the secondary side in order to achieve the best performance. Specifically, in order to reduce the search space, we have set the primary coil dimension to 75 cm × 75 cm, whereas the secondary side has a size of 50 cm × 50 cm. This difference makes the system less sensitive to coil misalignment.

Depending on the operational frequency, the coils should be built with Litz wire. Specifically, as shown in Table 3.2, Litz wire should have a strand gauge of AWG38 (diameter equal to 0.1007 mm). In the iterative algorithm, the potential wire sections

Table 6.3 Potential solutions derived from the design of the coils

Number of turns of the primary coil (N1)	Number of turns of the secondary coil (N2)	Cross section of the primary coil's wire (mm²)	Cross section of the primary coil's wire (mm²)	Efficiency	Copper (kg)
14	10	5.35	5.35	0.9893	2.972
14	10	5.35	8.51	0.9906	3.538
14	10	8.51	5.35	0.9920	4.161
14	10	8.51	8.51	0.9932	4.727
14	10	13.38	5.35	0.9936	5.993
14	10	13.38	8.51	0.9948	6.560

Table 6.4 Properties of the chosen solution

Property	L_1 (μH)	L_2 (μH)	M (μH)	q_1	q_2
Value	475	152	37	12.7	4.1

have been limited to those recommended by the manufacturers [10]. Additionally, solutions must avoid bifurcation and yield to an efficiency greater than 95%. For our example, the potential solutions are described in Table 6.3.

From this set, a primary coil composed of 14 turns and a secondary side composed of 10 turns are selected. The cross section of the wire for both coils is selected at 8.51 mm², given that the 5.35 mm² solution could be insufficient for the nominal current of the charger (particularly if a misalignment occurs), while the 13.38 mm² solution would increase the weight and the cost. The other properties of the chosen solution are defined in Table 6.4.

As previously defined, the compensation networks are mono-resonant with a Series-Series topology. The computation of the capacitors are based on the results of Eq. 4.11, which yields to:

$$C_1 = \frac{1}{L_1\omega_0^2} = \frac{1}{475 \cdot (2\pi \cdot 85000)^2} = 7.4nF \tag{6.1}$$

$$C_2 = \frac{1}{L_2\omega_0^2} = \frac{1}{152 \cdot (2\pi \cdot 85000)^2} = 23nF \tag{6.2}$$

6.2.4 Design of the Power Converters

The wireless charger has been designed with the goal of minimising its complexity. Thus, there are only three power converters in the system: the three-phase rectifier and the inverter on the primary side, and the rectifier on the secondary side.

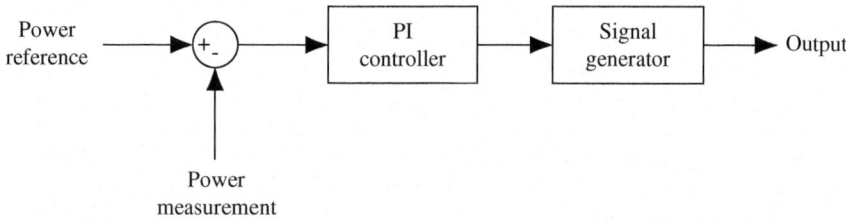

Fig. 6.4 Diagram of the phase-shift controller

Delivering 300 V to the battery makes the half-bridge single-phase rectifier non-feasible for the primary side without DC–DC converters and smaller coils on the secondary side. A full-bridge (single-phase or three-phase) rectifier may work properly, however. The output voltage may exceed 300 V, and so a transformer may be used in the rectifier input. Another alternative consists in regulating the voltage on the primary inverter. We have opted for using a full-bridge three-phase rectifier with a step-down transformer in the input. In this way, the input voltage of the inverter is close to that required by the battery. The system is also provided with isolation.

The primary inverter is key for controlling the amount of power transferred. In wireless chargers, its complexity is due to the high frequency at which the semiconductors must switch. The topology of this power converter will restrict the semiconductor devices that can be used and determine how the control will be implemented. Considering the analysis described in Chap. 5, we have opted for a full-bridge inverter composed of 4 MOSFETs. This is a well-known and highly efficient topology. A phase-shifting technique can be applied to this topology to control the power transferred. This adjustment is necessary to adapt the system configuration to any potential misalignment of the coils.

The last converter is the secondary rectifier, which must be capable of operating at 85 kHz. In our solution, we have selected a full-bridge non-controllable rectifier. This simple solution reduces the electronics and control components in the rectifier so that the cost of this part is lower than for the primary side.

6.2.5 Control Design

Based on the proposed structure of the wireless charger, the control algorithm can only be implemented in the primary inverter to regulate the power transfer. In particular, we have opted for a phase-shift control (Fig. 5.26). A PI-control (illustrated in Fig. 6.4) has been implemented to set the phase shift. The charging power is compared with the reference power (set initially) so that the phase shift is adjusted to equalise both magnitudes. The output of the PI controller configures the switching of the MOSFETs in the inverter.

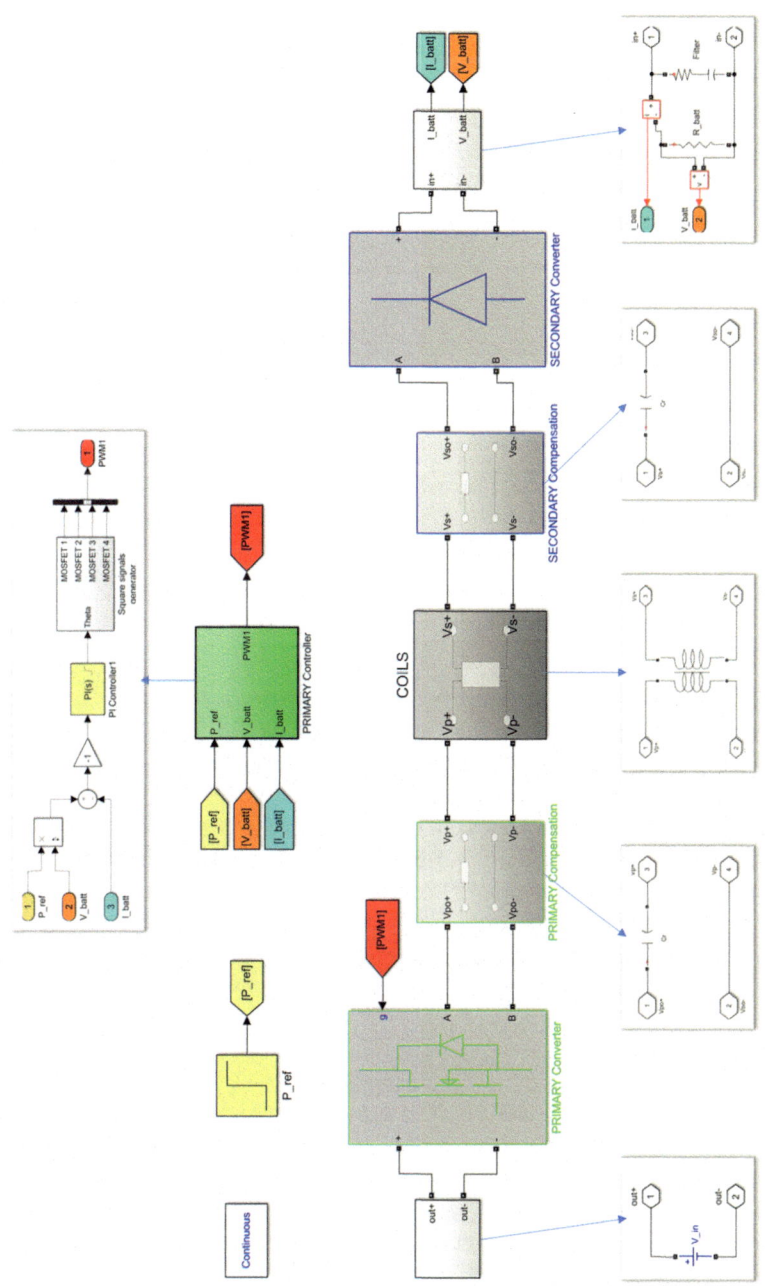

Fig. 6.5 Implementation of the prototype using the SIMULINK software

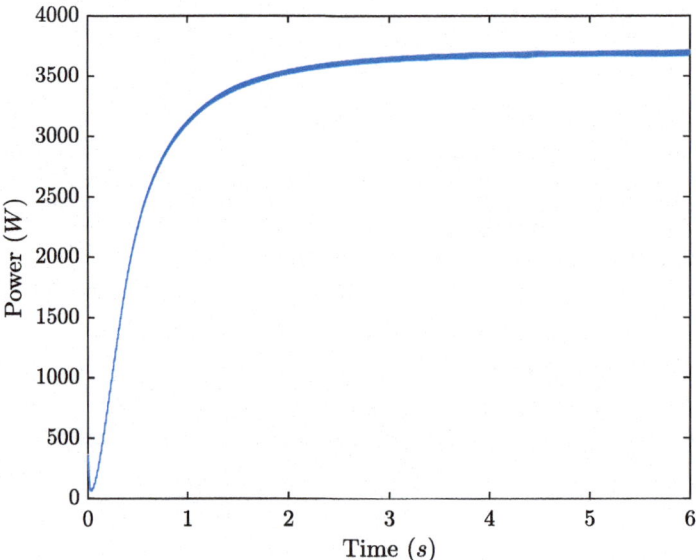

Fig. 6.6 Evolution of the power delivered to the battery

6.2.6 Software Tests

To validate the design, some software tests have been performed with the SIMULINK software [8]. The model used for this evaluation is included in Fig. 6.5.

Each part in the model is implemented with blocks. The contents of each block are included in the Fig. 6.5. The DC-link is modelled as a DC voltage source. The inverter is implemented with 4 MOSFETs in a full-bridge topology. Instead of using the simulink block for the inverter, constructing it from its semiconductors allows us to measure the electrical magnitudes in each device.

The control block is in the upper section of Fig. 6.5. As can be observed, this block has been implemented with a PI controller. The power delivered to the battery is computed based on the voltage and current measurements. The difference between this computation and the reference power is used to estimate the phase that will activate the MOSFETs.

Figure 6.6 shows how the power is stabilised at 3.7 kW. It is necessary to set a transient period (approximately 4 s) in order to avoid electrical damage to the components. The duration of the transient period can be adjusted with the parameters of the PI controller.

With the charger stabilised to provide 3.7 kW, we have analysed the output voltage and the current of the primary inverter (Fig. 6.7). The phase shifting has been correctly applied to reduce the power of the first harmonic of the output voltage. The power factor is close to unity as the current and voltage are nearly in phase practically.

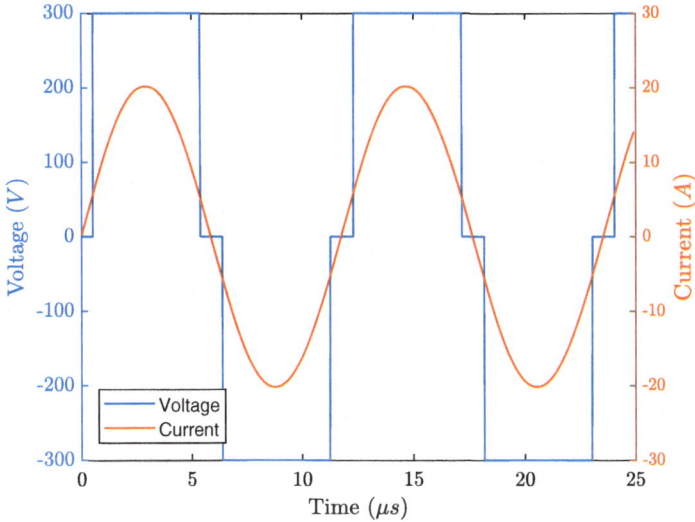

Fig. 6.7 Output voltage and current of the primary inverter

6.3 Prototype Implementation

Once the design has been verified in the software tests, the implementation is initiated. Figure 6.8 shows a photo of some of the components used for the prototype. In this figure, the components are numbered as follows:

1. Grid rectifier.
2. Primary DC-link. This includes capacitors in order to reduce the voltage ripple.
3. Primary controller.
4. Primary inverter.
5. Primary current sensor.
6. Primary compensation network.
7. Secondary compensation network.
8. Secondary rectifier.
9. Secondary controller.
10. Secondary current sensor.
11. Secondary DC-link. This includes capacitors in order to reduce the voltage ripple.

The following subsections explain our reasons for selecting these components and their configuration.

Fig. 6.8 Partial prototype implementation [4]

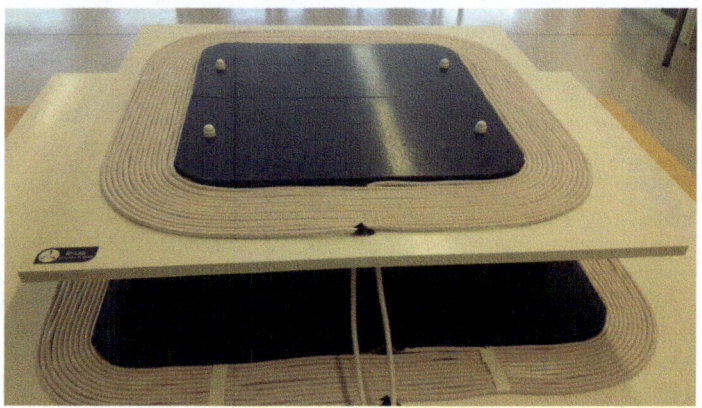

Fig. 6.9 Implemented coils

6.3.1 The Coupler

Taking into account the results of the algorithm used for the coil design, these components have been built accordingly. The two coils are built with Litz wire with a cross section equal to 8.51 mm^2. The wire is composed of 1.050 strands in a AWG-38 configuration. Although this type of wire is more expensive than copper, its use reduces the system losses at the operational frequency. Soldering the cables becomes more challenging, however.

Figure 6.9 shows the final implementation of the coils. They are built on wood, avoiding conductive materials that could alter the magnetic field. Other alternatives would be special plastic or methacrylate. When deciding the material for the base, it is necessary to analyse the temperature of the coils and whether these components should be protected from rain or dust. As the implemented prototype was built for research purposes only, these issues were not considered.

Fig. 6.10 Schematics of the
compensation networks

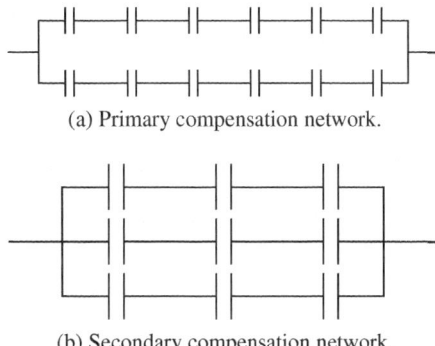

(a) Primary compensation network.

(b) Secondary compensation network.

6.3.2 Compensation Networks

The capacitors in the compensation networks should support a high variation rate of the voltage and current due to the operational frequency set in the wireless charger. Not all the capacitors comply with this requirement, and the polypropylene film capacitors are the most convenient for this kind of application [4, 7, 13].

The computation of the compensation networks has yielded to 7.4 and 23 nF on the primary and secondary side respectively. To implement these capacitors, it is necessary to associate commercial capacitors in series and/or in parallel. Additionally, series association is recommended to distribute the voltage among the various components, as the compensation network suffers from a high voltage.

In our prototype, we have implemented both compensation networks as shown in Fig. 6.10. The primary compensation network is the result of two parallel branches. In each branch there are 6 22-nF capacitors in series. For the secondary compensation network, three parallel branches have been used, each of which is composed of 3 22-nF capacitors in series. Although the combination of the secondary compensation network could be simplified to just one capacitor, the proposed implementation allows for a distribution of the voltage.

6.3.3 Power Converters

The three-phase full-bridge rectifier connected to the grid has been implemented with three SKKD 26/14 modules provided by SEMIKROM [12]. These modules constitute each branch in the rectifier and are composed of two diodes, which support up to 30 A as the mean nominal current and repetitive peak voltage of 1400 V. Their maximum forward voltage is 1.35 V. Figure 6.11 shows a photo of the three modules.

Fig. 6.11 Full bridge
rectifier used on the primary
side

Fig. 6.12 KIT8020CRD8FF1217P-
1 evaluation board

The primary inverter has been designed and implemented to support switching frequencies of 85 kHz. Due to this operation frequency, IGBT are not appropriate for the system and MOSFETs have been selected as switches. In particular, SiC MOSFETs are the ones best able to support the currents and voltages required by this present application. The use of MOSFETs requires the inclusion of additional electronics to activate the switches correctly and to protect them with isolation systems and snubbers. Heat sinks are also necessary for the range of power managed in this charger. To ease the implementation, we have opted for the CREE KIT8020CRD8FF1217P-1 evaluation board [1], which comprises all the components for the safe operation of a MOSFET leg. This includes the free-wheeling diodes C4D20120D [2]. This board is illustrated in Fig. 6.12.

The two boards are connected as shown in the diagram in Fig. 6.13.

The secondary rectifier has been implemented with the same evaluation boards, making use of the free-wheeling diodes. These diodes are Silicon Carbide Schottky, which means that their conduction losses are negligible. This kind of implementation will also make it easy to extend the prototype for bi-directional power transfer.

Fig. 6.13 H-bridge topology configuration using two KIT8020CRD8FF1217P-1 boards [1]

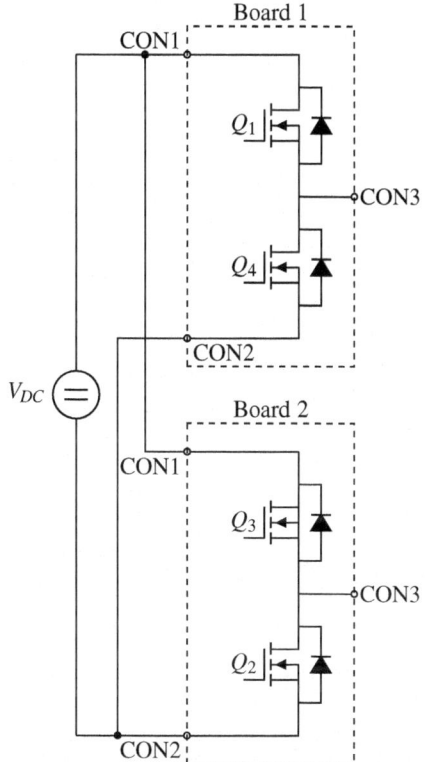

6.3.4 Controllers

The implementation of the control algorithm must be supported by a controller with the following features:

1. Capacity to generate a 85-kHz PWM signal.
2. High speed to process the instructions.
3. Real Time Operating System (RTOS).
4. Capacity to use measurements from sensors. In the event that analogue sensors are used, A/D converters are required to process the data.
5. Interface for wireless communication.

As a result of the study on commercial controllers that comply with these requirements, we have identified the Intel Edison, the BeagleBone Blue and the Arduino Yun as valid computer boards. We have opted for the Intel Edison Board because of its availability in our lab [6]. This is a development board with a two-core Intel Atom processor (its clock frequency is 500 MHz) and a 100-MHz Intel Quark microcontroller. It has 6 analogue inputs and a Bluetooth Low Energy interface. The price of the component is approximately 80.

Although this board already has a micro-controller, we have incorporated an additional micro-controller to generate the MOSFET activation signals. With this new component, dsPIC30F4011 by Microchip [9], the charger can be activated without the Intel Edison, which gives a clear advantage during the tests. These elements are assembled in a PCB as shown in Fig. 6.8 with the identifiers 3 and 9.

Data communication between the primary and secondary sides is achieved through the Intel Edison boards. These boards have an IEEE 802.11 interface and a Bluetooth Low Energy (BLE) interface. IEEE-802.11 link could be used for communication between the charger and the BMS. However, it is recommended that the communication between the two boards should be performed with BLE. This recommendation is based on the following advantages offered by this protocol:

- Reduced power consumption
- Data transfer at 24 Mbps.
- The maximum coverage range is 30 m, which is sufficient for our application.

The Intel Edison boards are programmed with the Python language to:

1. Implement the procedures to establish communication between the boards on the primary and secondary sides.
2. Develop MODBUS communication with other components of the system.
3. Process the signals derived from the measurements taken from the MODBUS or the sensors connected to the analogue inputs.
4. Compute the phase shift with a PI control.
5. Send the computed phase shift to the PIC so that they can generate the activation signals of the MOSFETs accordingly.

The user interface is a Linux-command window where the power can be adjusted. The sequence of operation of this controller is as follows:

1. Both Intel Edison boards initiate communication through the BLE interface. The board on the secondary side triggers the MODBUS communication with the other components, if required.
2. The board on the secondary side periodically checks whether the battery is ready to be charged. The board on the primary side remains in an idle state so that the MOSFET activation signals are not generated.
3. Once the battery is ready to be charged, the board on the secondary side captures the current and voltage of the battery through the sensors connected to its inputs AN0 and AN1.
4. From the power and voltage measurements taken in the battery, the board on the secondary side computes the active power delivered to the load.
5. The secondary board sends the computed power to the board on the primary side through the BLE channel.
6. The board on the primary side estimates the difference between the power being delivered to the battery and the target power. Based on the PI control, it computes the phase shifting to apply.
7. The obtained phase shift is sent from the Intel Edison to the PIC through a parallel bus.

8. The PIC generates four square waves to activate the MOSFETs in the primary inverter according to the phase shift.
9. This process is repeated until the battery receives the target power.

Alternatively, the PIC has been programmed using the MPLAB X IDE software by Microchip, which makes it possible to program using the C language.

6.3.5 Load Modelling

For the experimental tests, we have used a power resistance that models the battery. This element has a resistance of 25 Ω, which is an approximate value of:

$$R_L = \frac{V_{bat}^2}{P_{charge}} = \frac{300^2}{3700} = 24.3\Omega \tag{6.3}$$

Electronics charge could be an alternative for this load.

6.3.6 Summary of the Main Parameters of the Implemented Prototype

The next Table describes the electrical features of the main components used in our prototype (Table 6.5).

Table 6.5 Main electric components of the system

	Primary		Secondary	
Compensation network	C_1	7.4 nF	C_2	23 nF
Magnetic coupling				
Self-inductance	L_1	475 μH	L_2	152 μH
Mutual inductance	$M = 37\,\mu$H			
PI control				
Proportional constant	k_{p1}	0.00001	k_{p2}	–
Integral constant	k_{i1}	0.01	k_{i2}	–

6.4 Lab Tests

The experimental tests have been carried out increasing the delivered power progressively. The following Figures show the voltage and current for the input and output of the primary inverter and the secondary rectifier with the system working at maximum power. As was the case in the software tests, it can be observed that the charger is operating at ZPA, which increases the efficiency (Figs. 6.14 and 6.15).

Based on these measurements, the efficiency of the wireless power transfer can be computed. This metric is summarised in Table 6.6.

The efficiency is related to the losses in the primary inverter, the secondary rectifier, the coils and the compensation networks. The grid rectifier also suffers losses, but it will not be included in this analysis. Efficiency can be computed as shown in Eq. 6.4.

$$\eta = \frac{P_{grid} - L_{inv} - L_{rec} - L_{coils} - L_{match}}{P_{grid}} \tag{6.4}$$

The losses of the inverter are due to conduction and switching losses. Based on measurements taken, the losses equal the difference between the input power ($P_{inv_{input}}$) and the output power ($P_{inv_{output}}$) of the inverter as presented in Eq. 6.5.

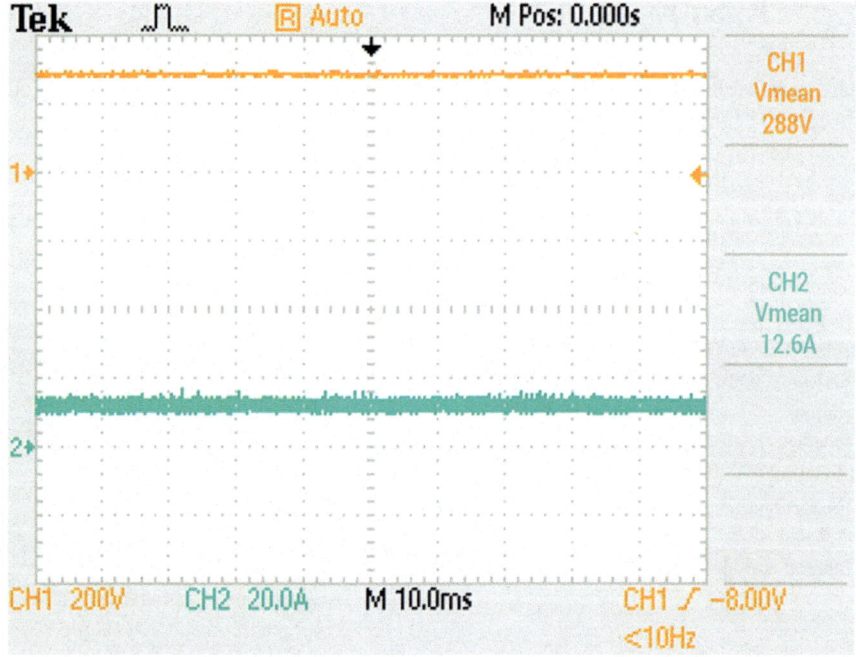

Fig. 6.14 Input of the primary inverter

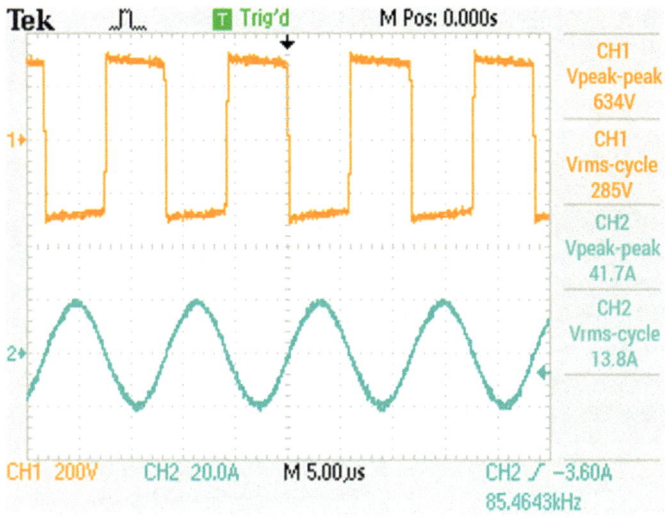

Fig. 6.15 Output of the primary inverter

Table 6.6 Electrical signals measured in the prototype

Electrical signals	Value
$V_{inv_{input}}(V)$	288
$V_{inv_{output}}(V)$	290
$I_{inv_{input}}(A)$	12.56
$I_{inv_{output}}(A)$	13.78
$V_{rec_{input}}(V)$	285
$V_{rec_{output}}(V)$	288
$I_{rec_{input}}(A)$	13.74
$I_{rec_{output}}(A)$	12.16

$$L_{inv} = P_{inv_{input}} - P_{inv_{output}} \qquad (6.5)$$

where $P_{inv_{input}}$ is calculated with Eq. 6.6 and $P_{inv_{output}}$ with Eq. 6.7. For the latter, as the voltage wave corresponds to a square signal, it is necessary to proceed with the harmonic analysis (Figs. 6.16 and 6.17).

$$P_{inv_{input}} = V_{inv_{input}} \cdot I_{inv_{input}} \qquad (6.6)$$

$$P_{inv_{output}} = \frac{V_{inv_{output}}^{1} \cdot I_{inv_{output}}}{2} = \frac{4 \cdot V_{inv_{output}} \cdot I_{inv_{output}}}{\pi \cdot \sqrt{2}} \qquad (6.7)$$

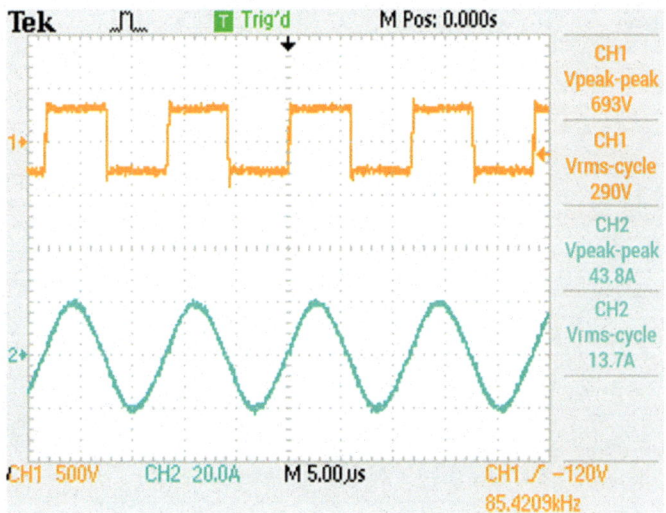

Fig. 6.16 Input of the secondary rectifier

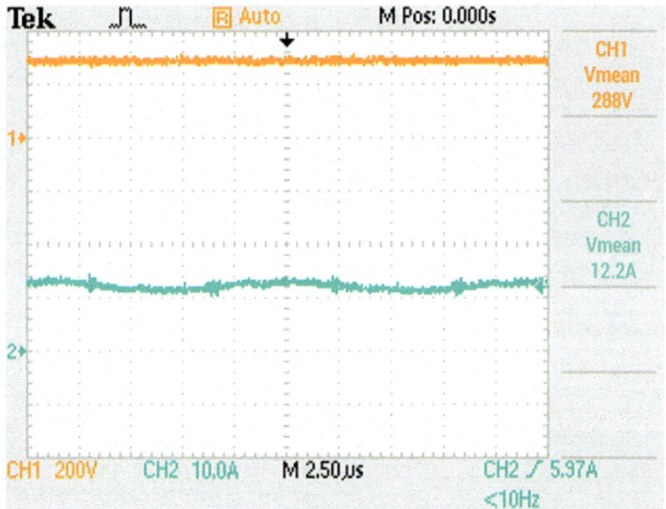

Fig. 6.17 Output of the secondary rectifier

The losses of the rectifier have the same cause as those of the inverter and are therefore computed similarly. In this case, the losses of the rectifier are the difference between the input power ($P_{rec_{input}}$) and the output power ($P_{rec_{output}}$) of the rectifier and yields to the following expression:

$$L_{rec} = P_{rec_{input}} - P_{rec_{output}} \tag{6.8}$$

Table 6.7 Losses estimated from the electrical signals measured in the prototype

Losses	
$L_{inv}^{ch}(W)$	20
$L_{coils} + L_{match}(W)$	73
$L_{rec}^{ch}(W)$	23

where $P_{rec_{input}}$ is calculated with Eq. 6.9 and $P_{rec_{output}}$ with Eq. 6.10. For the former, the voltage wave also corresponds to a square signal; consequently it is necessary to apply the harmonic analysis.

$$P_{rec_{input}} = \frac{4 \cdot V_{rec_{input}} \cdot I_{rec_{input}}}{\pi \cdot \sqrt{2}} \tag{6.9}$$

$$P_{rec_{output}} = V_{rec_{output}} \cdot I_{rec_{output}} \tag{6.10}$$

Finally, the losses of the coils (L_{coils}) and the compensation networks (L_{match}) are basically due to conduction losses. Having the output power of the inverter ($P_{inv_{output}}$) and the input power of the rectifier ($P_{rec_{input}}$), these losses are computed as::

$$L_{coils} + L_{match} = P_{rec_{input}} - P_{rec_{output}} \tag{6.11}$$

Using the measurements and the previous equation, we have computed the losses in the system, which are specified in Table 6.7. The efficiency of the system is approximately 96.9%.

References

1. CREE: Silicon Carbide MOSFET Evaluation Kit KIT8020CRD8FF1217P-1 Datasheet (2014). http://go.pardot.com/l/101562/2015-08-10/9z1/101562/854/KIT8020_CRD_8FF1217_1. pdf
2. CREE: C4D20120D Datasheet (2016). https://www.wolfspeed.com/media/downloads/106/ C4D20120D.pdf
3. Fischer, H.M., Dorn, L.: Voltage Casses for Electric Mobility. Tech. rep., ZVEI - German Electrical and Electronic Manufacturers Association (2013). https://www.zvei.org/fileadmin/ user_upload/Presse_und_Medien/Publikationen/2014/april/Voltage_Classes_for_Electric_ Mobility/Voltage_Classes_for_Electric_Mobility.pdf
4. González-González, J.M., Triviño-Cabrera, A., Aguado, J.A., González-González, J.M., Triviño-Cabrera, A., Aguado, J.A.: Design and validation of a control algorithm for a SAE J2954-compliant wireless charger to guarantee the operational electrical constraints. Energies 11(3), 604 (2018). https://doi.org/10.3390/en11030604, http://www.mdpi.com/1996-1073/11/ 3/604

5. Iclodean, C., Varga, B., Burnete, N., Cimerdean, D., Jurchiş, B.: Comparison of different battery
    types for electric vehicles. IOP Conf. Ser.: Mater. Sci. Eng. **252**(1), 012,058 (2017). https://doi.
    org/10.1088/1757-899X/252/1/012058, http://stacks.iop.org/1757-899X/252/i=1/a=012058?
    key=crossref.39eab45305798f03f8f65392179a6747
 6. Intel: Intel® Edison Compute Module: Hardware Guide (2016). https://www.intel.com/
    content/dam/support/us/en/documents/edison/sb/edison-module_HG_331189.pdf
 7. Kan, T., Nguyen, T.D., White, J.C., Malhan, R.K., Mi, C.C.: A new integration method for
    an electric vehicle wireless charging system using LCC compensation topology: analysis and
    design. IEEE Trans. Power Electron. **32**(2), 1638–1650 (2017). https://doi.org/10.1109/TPEL.
    2016.2552060, http://ieeexplore.ieee.org/document/7448975/
 8. MathWorks: Simulink - Simulation and Model-Based Design - MATLAB & Simulink. https://
    www.mathworks.com/products/simulink.html
 9. Microchip: dsPIC30F4011/4012 datasheet (2005)
10. New England Wire Technologies: Litz Wire: Technical Information. http://www.litzwire.com/
    nepdfs/Litz_Technical.pdf
11. SAE International: Wireless Power Transfer for Light-Duty Plug-In/Electric Vehicles and
    Alignment Methodology (SAE TIR J2954) (2017)
12. SEMIKRON: SKKD 26 Datasheet (2016)
13. Zhang, H., Lu, F., Hofmann, H., Liu, W., Mi, C.: A 4-plate compact capacitive coupler design
    and LCL-compensated topology for capacitive power transfer in electric vehicle charging
    applications. IEEE Trans. Power Electron. 1–1 (2016). https://doi.org/10.1109/TPEL.2016.
    2520963, http://ieeexplore.ieee.org/document/7390090/

Appendix
Software Tools

This appendix contains some of the software referenced in the text. The developed software helps readers to perform some tests in order to initiate the design of EV wireless chargers. Its execution requires the use of MATLAB.

In order to help readers manage the software, a user-friendly interface has been developed using the Graphical User Interface Development Environment (GUIDE), a MATLAB module which makes it easier for users to control the software applications.

To execute the interface, the user has to select the software folder in the "Current Folder" layout of MATLAB and right-click on the file "main.m". Finally, after clicking on the "Run" option, the main screen of the graphical interface will open (Figs. A.1 and A.2).

Fig. A.1 Main screen of included software

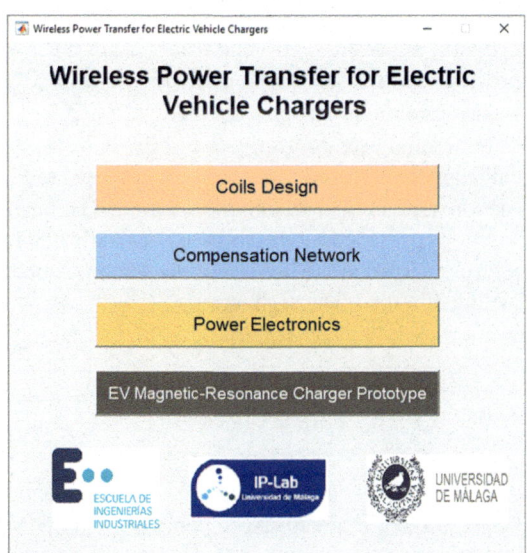

© Springer Nature Switzerland AG 2020
A. Triviño-Cabrera et al., *Wireless Power Transfer for Electric Vehicles:
Foundations and Design Approach*, Power Systems,
https://doi.org/10.1007/978-3-030-26706-3

The main screen allows the user to select the option that she wants to test. These options include tests related to the design of the coil, the compensation network, the power electronics and the magnetic resonance prototype developed in Chap. 6. The content is explained in the following sections.

A.1   Coil Design

The coil design depends on several factors such as the geometry and the compensation network that will be used. In our software, we have included algorithms for rectangular coils. The "Coil Design" screen allows to select two options:

- Design of Coupled Coils.
- L-Calculation.

The last option presents a script with the basis of the theoretical calculation of L parameter of rectangular coils, while the former open a new screen (Fig. A.3).

The "Design of Coupled Coils" screen allows users to design a coils for wireless chargers, providing parameters as numbers of turns of each coils, areas, efficiency, etc. In order to achieve this solution, users have to introduce some design data, such as maximum number of turns, size of coils, gap between coils, etc. (Fig. A.4).

A.2   Compensation Networks

Although the "Coil Design" section helps the user to calculate the compensation network, a specific section on compensation networks has been included. Unlike the previous section, this option allows the user to work not only with the four basic mono-resonant topologies but also with LCC.

When the user clicks on one of the buttons, the interface shows a new screen to calculate the selected topology. As presented in Fig. A.5, the screen is divided in two sections: the input data and the solution. The labels of both the data required and the solutions provided by the software are indicated in the diagram of the compensation network shown at the bottom of the screen. A button to open the script has also been included in the lower right corner.

A.3   Power Electronics

The "Power Electronics" section includes models of the converters presented in Chap. 5. Each converter is accompanied by SIMULINK models which allow the user to test it. The implemented models include:

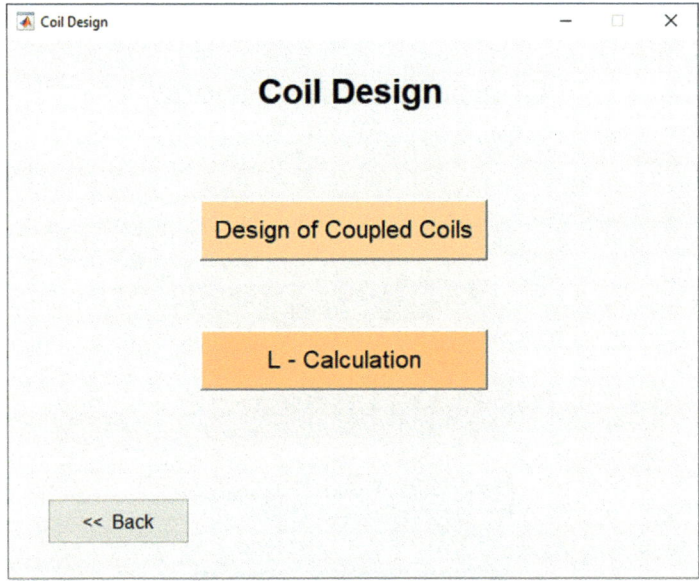

Fig. A.2 Coil design screen of included software

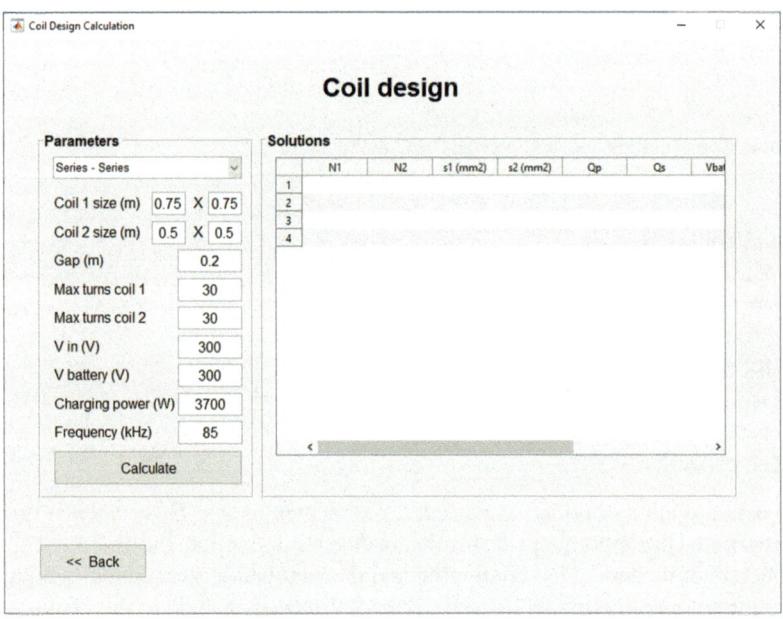

Fig. A.3 "Design of coupled coils" screen

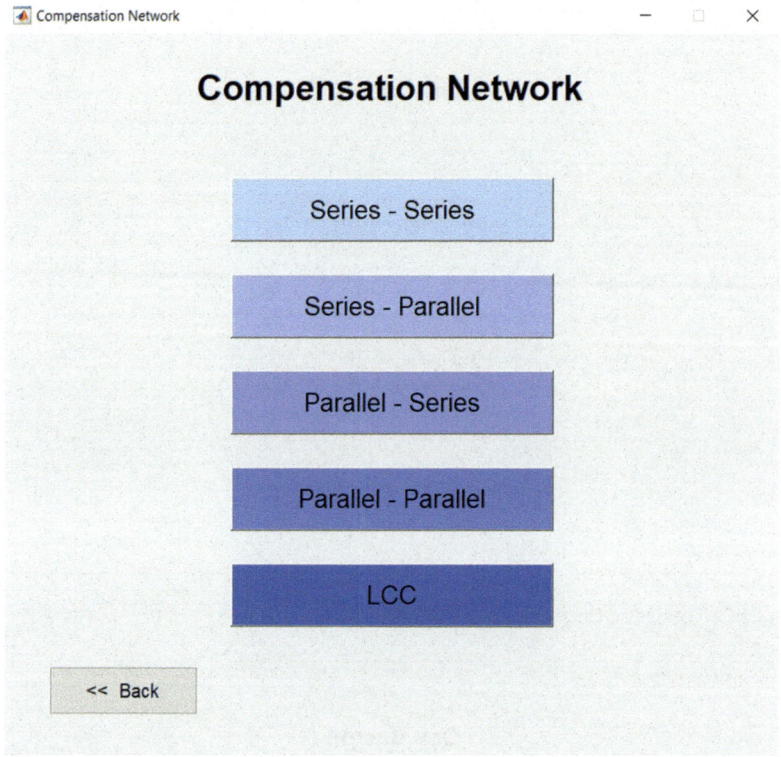

Fig. A.4 Compensation network screen of included software

- Single-phase rectifier.
- 3-phase rectifier.
- Controlled rectifier.
- Half-bridge inverter.
- Full-bridge inverter.
- Phase-shifted inverter.
- Boost converter.
- Buck converter.

When a topology option is selected, a new screen opens. This screen is divided into two parts: the options and the results. In the options section, the different editable parameters of the model are configured and the simulation is executed. In addition, an option to open the model directly in SIMULINK is included. The results show representative plots of the operation of the model (Figs. A.6 and A.7).

All SIMULINK models included follow a similar structure, with power electronics, a basic control, a filter and a load. Next to each one is a block of data output, a real-time plot and an external plot to visualise further SIMULINK data (Fig. A.8).

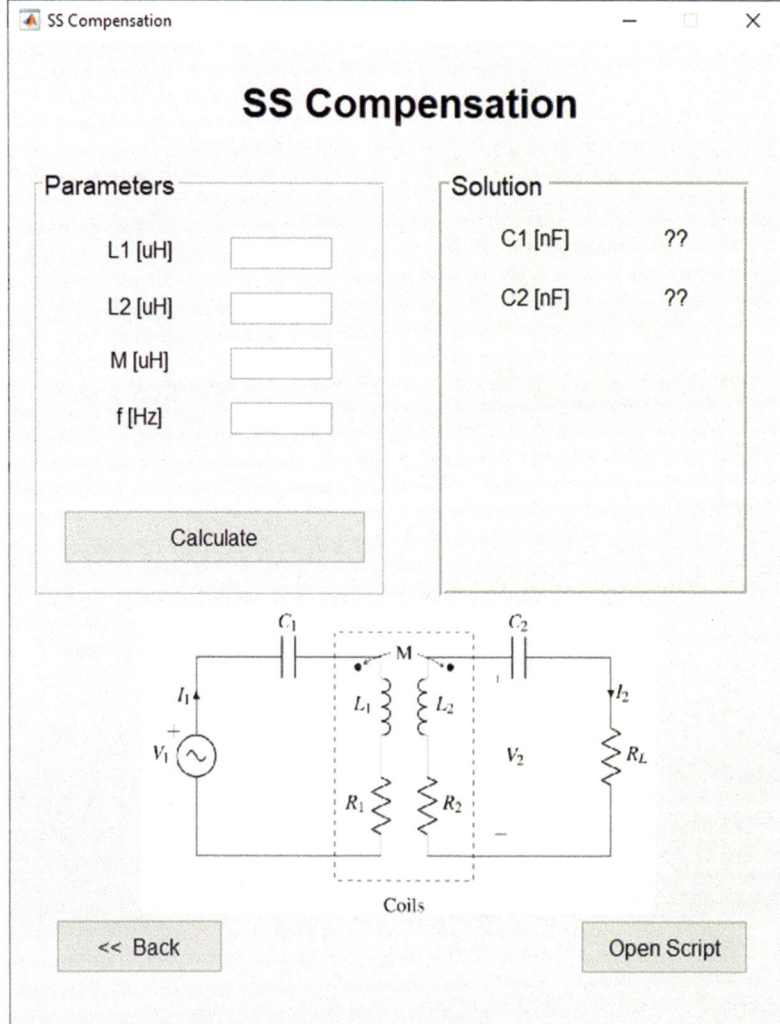

Fig. A.5 SS compensation screen of included software

A.4   EV Magnetic Resonance Charger Prototype

This option of the software focuses on the Magnetic-Resonant Charger Prototype developed in Chap. 6. A SIMULINK model is included to simulate the prototype, allowing the users to configure different parameters as coils, compensation networks, control.

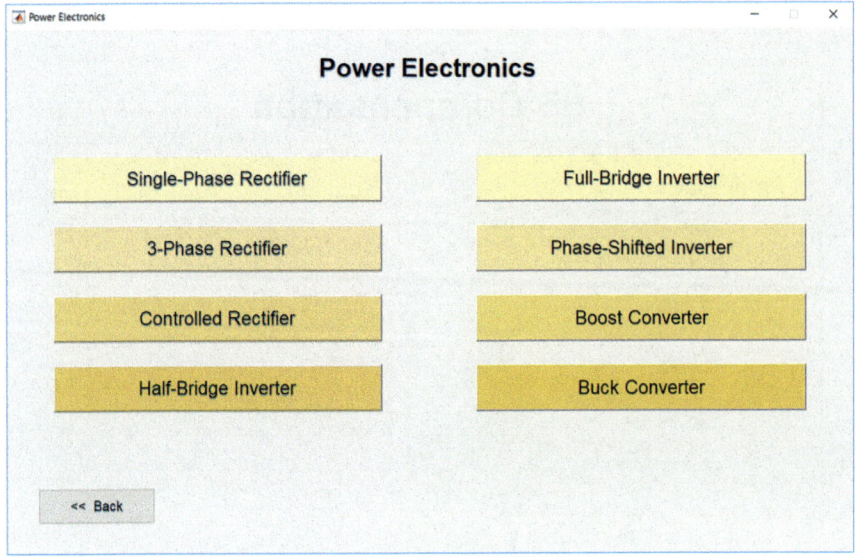

Fig. A.6 Power electronics screen of included software

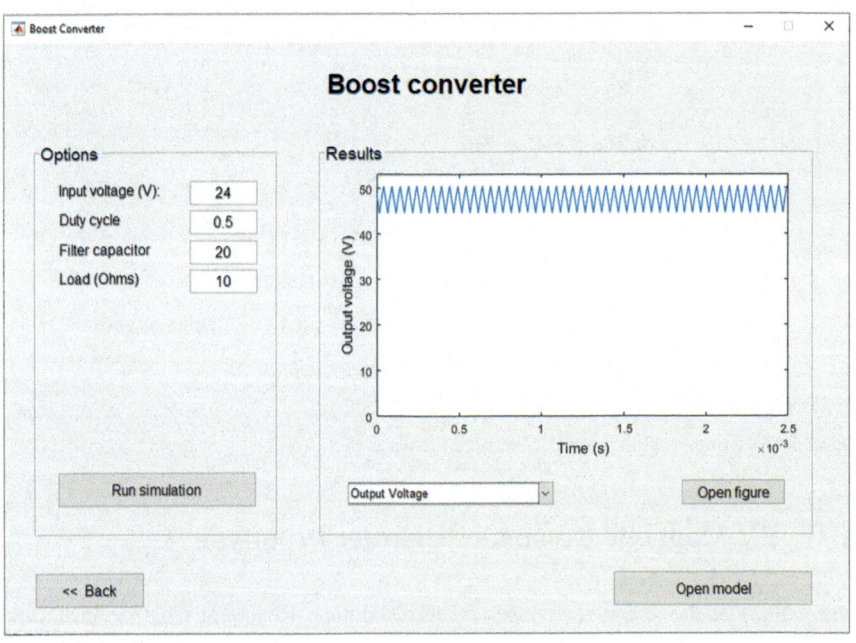

Fig. A.7 Boost converter screen of included software

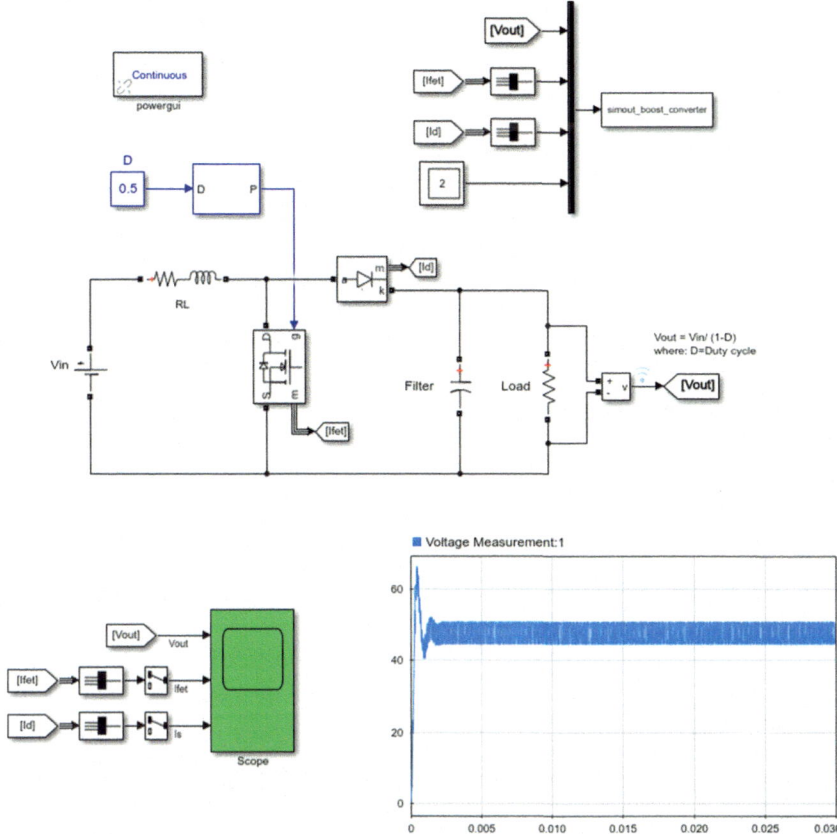

Fig. A.8 Boost converter SIMULINK model

Index

A
AC/DC converter, 108, 118, 120, 124

B
Batteries, 20
Battery management system, 23, 146
Bi-directional wireless power transfer, 31
Bifurcation phenomenon, 71, 72, 75, 78, 79, 82, 85, 87, 89, 132
Bipolar coils, 49
Boost converter, 119
Buck converter, 119

C
Capacitive WPT, 8
Car manufacturers, 36
Circular coil, 45, 46
Compensation network, 69, 133, 137
Controlled rectifier, 109
Controller, 138, 145

D
DC/AC converter, 113
DC/DC converter, 119, 125
DD coils, 48
DDQ coils, 48
Dynamic charging, 27, 30, 49

F
Full-bridge inverter, 115, 138, 144
Full-bridge rectifier, 108, 138, 143

H
Half-bridge inverter, 114
Half-bridge rectifier, 108

I
Inductive WPT, 5
International Electrotechnical Commission (IEC), 33
International Organization for Standardization (ISO), 33
Inverter, 113

L
LCC compensation, 92, 133
LCL compensation, 90
Lead-acid battery, 21
LFO battery, 22
Li-ion battery, 21
Litz wire, 50, 136

M
Magnetic resonance WPT, 7, 69
Market outlook, 34
Microwave power transfer, 10
Misalignment, coil, 47, 75, 88, 97
Mono-resonant compensation, 7, 72
Multi-level inverters, 117
Multi-resonant compensation, 7, 89

N
NiMH battery, 21
NMC battery, 21
NVA battery, 21

© Springer Nature Switzerland AG 2020
A. Triviño-Cabrera et al., *Wireless Power Transfer for Electric Vehicles: Foundations and Design Approach*, Power Systems,
https://doi.org/10.1007/978-3-030-26706-3

O
Optical WPT, 12

P
Parallel–parallel compensation, 84
Parallel–series compensation, 79
Phase-locked loop, 122
Phase-shifting, 115, 138
Power factor corrector, 110
Pulse width modulation, 113, 121, 145

R
Resonant WPT, 7

S
Series–parallel compensation, 77

Series-series compensation, 73, 133, 136, 137
SiC MOSFET, 103
Society of Automotive Engineers (SAE), 32
Static charging, 29
Stationary charging, 30
Strongly coupled magnetic resonance, 9

T
Three-phase rectifier, 108, 138

V
Vehicle to grid, 31, 101, 120

Z
ZEBRA battery, 21

Printed by Printforce, the Netherlands